Advances in Anatomy Embryology and Cell Biology

Vol. 149

AF719131

Book Stacks
QL801.E67
Advances in anatomy,
embryology and cell biolo
; v.
ct/t

Editors
F. Beck, Melbourne D. Brown, Charlestown
B. Christ, Freiburg W. Kriz, Heidelberg
E. Marani, Leiden R. Putz, München
Y. Sano, Kyoto T. H. Schiebler, Würzburg
K. Zilles, Düsseldorf

Springer

Berlin
Heidelberg
New York
Barcelona
Hong Kong
London
Milan
Paris
Singapore
Tokyo

M. Klima

Development of the Cetacean Nasal Skull

With 68 Figures and 1 Table

Springer

Prof. Dr. M. Klima
Dr. Senckenbergische Anatomie, Institut III
Johann Wolfgang Goethe-Universität
Theodor-Stern-Kai 7
60590 Frankfurt am Main, Germany

QL 801 .E67 v. 149

ISBN 3-540-64996-4 Springer-Verlag Berlin Heidelberg New York

Library of Congress-Catalog-Card-Number

Klima, Milan. Development of the cetacean nasal skull / M. Klima. p. cm. — (Advances in anatomy, embryology, and cell biology, vol. 149) Includes bibliographical references and index. ISBN 3-540-64996-4 (softcover : alk. paper) 1. Cetacea—Morphogenesis. 2. Cetacea—Development. 3. Nasal capsule. I. Title. II. Series. [DNLM: 1. Cetacea—embryology. 2. Nose—embryology. W1AD433K v. 149 1998] QL801.E67 vol. 149 [QL737.C4] 571 s—dc21 [571.8'6195] DNLM/DLC for Libaray of Congress

This work is subject to copyright. All rights are reserved, whether the whole or part of the material is concerned, specifically the rights of translation, reprinting, reuse of illustrations, recitation, broadcasting, reproduction on microfilms or in any other way, and storage in data banks. Duplication of this publication or parts thereof is permitted only under the provisions of the German Copyright Law of September 9, 1965, in its current version, and permission for use must always be obtained from Springer-Verlag. Violations are liable for prosecution under the German Copyright Law.

© Springer-Verlag Berlin Heidelberg 1999
Printed in Germany

The use of general descriptive names, registered names, trademarks, etc. in this publication does not imply, even in the absence of a specific statement, that such names are exempt from the relevant protective laws and regulations and therefore free for general use.

Product liability: The publishers cannot guarantee the accuracy of any information about dosage and application contained in this book. In every individual case the user must check such information by consulting the relevant literature.

Production: PRO-EDIT GmbH, D-69126 Heidelberg
SPIN: 10676138 27/3136-5 4 3 2 1 0 – Printed on acid-free paper

Contents

Abbreviations

ca	cupula nasi anterior
cad	cupula nasi anterior dextra
cas	cupula nasi anterior sinistra
cdnp	cartilago ductus nasopalatini
con	commissura orbitonasalis
conl	commissura orbitonasalis, lower branch
conu	commissura orbitonasalis, upper branch
cp	cupula nasi posterior
cps	cartilago paraseptalis
crsc	crista semicircularis
dnp	ductus nasopalatinus
et	ethmoidale
ett1	ethmoturbinale I
ett2	ethmoturbinale II
ett3	ethmoturbinale III
fo	foramen occipitale
fol	foramen olfactorium
fon	fissura orbitonasalis
fr	frontale
frt	frontoturbinale
fs	anterior side wall structures (front side structures)
hs	posterior side wall structures (hind side structures)
jg	jugale
lcr	lamina cribrosa
lta	lamina transversalis anterior
Mo	Meckelian ossicle
mrf	fossa mesorostralis
ms	median structures
mx	maxillare
mxt	maxilloturbinale
N	nasal passage
ns	nasale
nt	nasoturbinale
oc	occipitale
pal	palatinum
par	parietale
plv	processus lateralis ventralis

pmx	praemaxillare
pn	paries nasi
prmx	processus maxillaris posterior
prpn	processus paranasalis
psp	processus paraseptalis posterior
pt	pterygoid
rett	recessus ethmoturbinalis
rfrt	recessus frontoturbinalis
rmx	recessus maxillaris
rn	rostrum nasi
sio	septum interorbitale
sn	septum nasi
spm	spina mesethmoidalis
sq	squamosum
ss	sulcus supraseptalis
tn	tectum nasi
ty	tympanicum
vom	vomer

1 Introduction

> The whale has no external nose; and that what nose he has – his spout-hole – is on the top of his head ...
> Herman Melville
> Moby Dick: or, The White Whale (1851)

Although the fact that cetaceans are not fish but mammals has been known at least since Aristotles time, and although whales have been mercilessly hunted for centuries, with thousands upon thousands having been observed and dissected, very little was known for a long time about their life habits, as well as about their anatomy. Thus, until the late 19th century there were only vague ideas about the way they breathe. It was far from clear that the comparatively small protuberance that in most cases was the only part of the whale's giant body which emerged from the water, emitting a geyser of steam, should in fact be the true nose. It was thought to be a "spout-hole" from which, according to the ancient illustrations, "the swallowed water" sprang like a true fountain (Fig. 1). The true nose was thought to be in the small pit at the tip of the rostrum, bearing tactile hairs in some cetaceans (see Lacepède 1804; Von Baer 1826; Kükenthal 1893).

The fact that the cetacean nose moved, in the course of evolution, from the tip of the rostrum up to the vertex of the head, is among the most perfect of adaptations to aquatic life. In this and many other special adaptations of their morphology and physiology, cetaceans surpass most primarily aquatic animals even though they themselves have developed from land mammals that breathe with lungs, and have only secondarily conquered the aquatic environment. To a certain extent, cetaceans can be considered to be the most successful group of aquatic animals of all time.

Conclusive paleontological evidence shows the way in which the nasal openings were moved in the course of phylogeny (see Kellogg 1928; Slijper 1962; Gaskin 1976; Oelschläger 1978, 1987, 1990; Moore 1981). That this evolutionary process is repeated in a way during ontogeny became obvious through external observations on embryos and fetuses (Kükenthal 1893). At the earliest embryonic stages the nasal openings are still situated at the rostral tip like those of land mammals; they are gradually shifted more and more towards the vertex of the head at the older stages. At the same time, a long rostrum with narrow jaws develops. Until recently, practically nothing was known about the morphogenetic processes concealed in this metamorphosis, about what cranial structures take part in it, and about the exact way in which the cetacean skull becomes transformed during embryogeny.

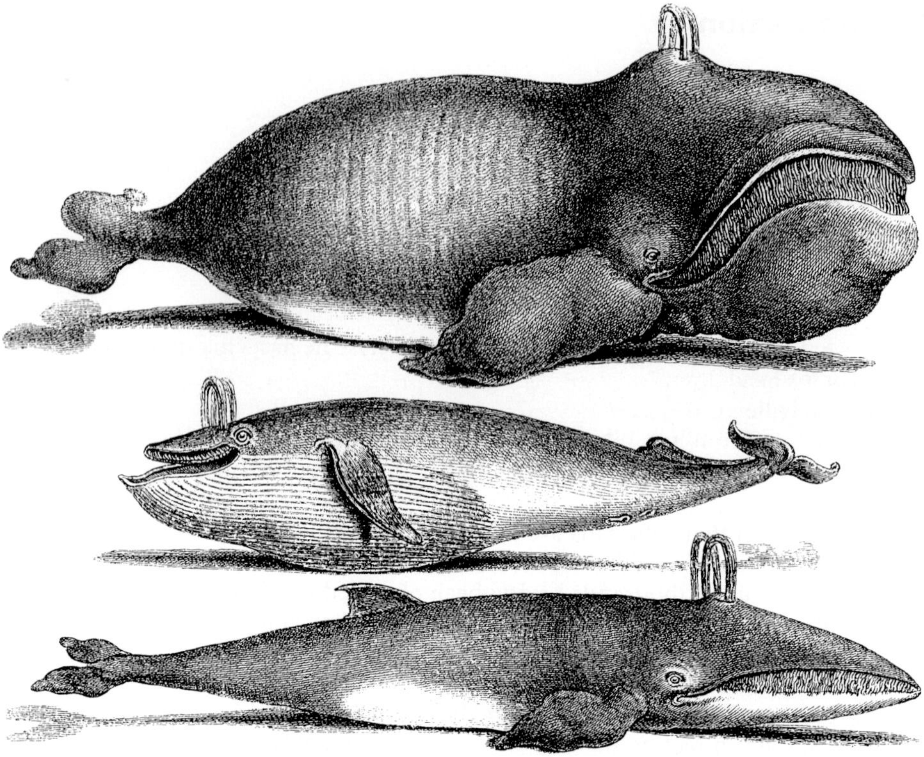

Fig. 1. In cetaceans the nostrils lie not on the tip of the snout but on the highest point of the whole huge cetacean body. The "spout-hole" on the top of their head is really the true nose – a fact unknown for a long time. From top to bottom: Bowhead whale, *Balaena mysticetus*; minke whale, *Balaenoptera acutorostrata*; and fin whale, *Balaenoptera physalus*. An ancient illustration from Lacepède (1804)

There are only a few papers concerning the morphogeny of the skull and the nasal skull of cetaceans. The oldest useful paper in this line is that by Kükenthal (1893). At the beginning of the 20th century, there followed the more detailed papers by De Burlet (1913a,b, 1914a,b, 1916) and Honigmann (1915,1917) which, at that time, were fundamental for the topic under study. They were followed by several minor papers (Schreiber 1916; Schulte 1916; Ridewood 1922; Eales 1950) and a comparative summary (De Beer 1937), which, however, contain no major new contributions to the topic. It was not until the 1990s that the topic was again tackled from new angles and quite new items of knowledge were obtained. This was due to the possibility of studying the morphogenesis of the nasal structures not only at individual stages, as was previously the case, but in a continuous developmental series of embryos and fetuses, first in *Phocoena phocoena* (Klima and Van Bree 1985) and later in *Physeter catodon* (Behrmann and Klima 1985; Klima et al. 1986; Klima 1987a, 1990a).

The present paper is a purposeful continuation of the latter papers. An attempt is made at clarifying the following complex of problems:

1. To provide a detailed description of the individual embryonic structures of the nasal skull in cetaceans

2. To establish the ontogenetic transformation and translocation processes of the nasal structures
3. To compare the embryonic nasal structures of the individual representatives of cetaceans mutually as well as with those of representatives of land mammals
4. To follow up the subsequent development of these structures up to the adult age and to clarify the homology of several questionable structures in the adult cetacean skull
5 To examine several functional connections, and above all, to clarify the consequences for the mode of surfacing and to make a comparison with other vertebrates that have perfectly adapted to the aquatic way of life, such as the Sirenia, among mammals, and the Ichthyosauria, among reptiles
6 To find out whether the morphogeny of the nasal structures can provide useful criteria for the phylogeny and systematics of Cetacea and, in a positive case, to establish to what extent the present systematic classification of Cetacea could thereby be changed.

Despite all endeavors, the results presented here are imperfect and incomplete. This is chiefly due to the fact that even the underlying material is incomplete. From among the 80 or so recent species of Cetacea, only 13 have been available, and two of the most important families, viz., Platanistidae and Ziphiidae, are not represented in our examinations. At the same time, the material examined is practically the entire embryological material of Cetacea available in all collections worldwide. Also, the material can hardly be expected to increase in the near future, since practically all cetacean species are immediately endangered by extinction and are subject to rigorous protection. To my knowledge, none of the world's research institutes or museums possess additional preserved materials of cetacean embryos. Thus, the present contribution, even though incomplete, is probably the most thorough treatment of the topic for some time to come. Hopefully, this paper may also emphasize the necessity of deepening and enlarging our general knowledge of this highly interesting mammal group while there still is an opportunity to do so.

2 Materials and Methods

For the present paper a total number of 62 embryos and fetuses from ten cetacean species were investigated (Table 1). Several other species and specimens examined for the present study are not mentioned in Table 1 or in the subsequent chapters. These include, on the one hand, the very early embryonic stages, in which the nasal structures have not yet been quite clearly developed (*Eschrichtius robustus*, two specimens, Seattle collection), and on the other, species of a genus in which another closely related species has been examined, yielding similar results (*Stenella coeruleoalba*, 11 specimens, collection Seattle and collection Tokyo, and *Stenella longirostris*, six specimens, Frankfurt collection). One species is additionally included in a detailed comparison, although no material of it was available, but detailed data were found in the literature (*Globicephala melas*, according to the papers by Schreiber 1916 and De Beer 1937). The most valuable data were obtained on those species for which a continuous developmental series of several specimens was available. This was the case for *Stenella attenuata, Delphinus delphis, Phocoena phocoena* and *Physeter catodon*. Besides, there was an opportunity for another direct comparison with a total of 11 embryos and fetuses of four species, reported in the earlier literature (Kükenthal 1893; Boenninghaus 1903; De Burlet 1913a,b, 1914a,b, 1916; Honigmann 1917). However, these four species are represented in our own material as well, and therefore, the number of species examined has not been increased. Thus, the total is 92 specimens of 13 species of Cetacea, evaluated for the purpose of the present paper.

The underlying material comes from the following collections: Frankfurt (Dr. Senckenbergische Anatomie, Johann Wolfgang Goethe University, Frankfurt/M., Germany), Seattle (National Marine Fisheries Service, Mammal Laboratory, Seattle, WA, USA), Tokyo (Department of Anatomy, Faculty of Medicine, University of Tokyo, Hongo, Tokyo, Japan) and Utrecht (Hubrecht Laboratory, International Embryological Institute, Utrecht, The Netherlands).

As regards the methods of taking measurements, the size of the embryos and fetuses is given in millimeters for the total length (TL), measured as the dorsal curved length from tip of snout to tip of tail, and for the crown-rump length (CR) measured as the straight line connecting the anterior top of head with the rounded posterior end of body or tail. The CR value can be somewhat more precisely determined at smaller stages up to about 50 mm TL, and the TL, at the bigger ones. Since most embryos and fetuses examined in this study belonged to the bigger stages, only the TL value has consequently been given throughout the paper, the CR values being additionally mentioned only in the list of materials.

The embryos and fetuses were sectioned into a series of microscopic slides following customary histological methods, except for the large fetuses which were examined by

Table 1. List of materials

Size in mm		Sectioned	Collections	Labeling	First reported
TL	CR				
Spotted dolphin, *Stenella attenuata*					
21.0	7.5	Transverse	Frankfurt	K 13	Present report
22.0	8.6	Transverse	Frankfurt	K 16	Present report
25.0	9.2	Sagittal	Frankfurt	K 12	Present report
39.0	14.8	Sagittal	Frankfurt	K 17	Klima (1987)
43.0	18.3	Transverse	Frankfurt	K 22	Klima (1987)
44.5	19.7	Sagittal	Frankfurt	K 23	Present report
57.0	25.5	Sagittal	Frankfurt	K 18	Klima (1987)
71.0	32.5	Sagittal	Frankfurt	K 24	Present report
76.0	30.0	Transverse	Frankfurt	K 14	Present report
85.0	38.5	Sagittal	Frankfurt	K 15	Klima (1987)
89.0	43.5	Horizontal	Frankfurt	K 25	Present report
115.0	58.5	Transverse	Frankfurt	K 27	Present report
153.0	82.0	Sagittal	Frankfurt	K 28	Present report
225.0	112.0	Transverse	Frankfurt	K 35	Present report
Common dolphin, *Delphinus delphis*					
24.0	7.6	Horizontal	Frankfurt	K 55	Present report
39.5	13.0	Horizontal	Frankfurt	K 56	Present report
42.0	18.7	Sagittal	Frankfurt	K 57	Present report
44.0	20.4	Horizontal	Frankfurt	K 53	Present report
46.0	21.5	Sagittal	Frankfurt	K 58	Present report
82.0	34.3	Transverse	Frankfurt	K 49	Present report
101.0	40.5	Sagittal	Frankfurt	K 50	Present report
113.5	49.5	Sagittal	Frankfurt	K 59	Present report
131.0	58.5	Transverse	Frankfurt	K 51	Present report
White-beaked dolphin, *Lagenorhynchus albirostris*					
107.0	46.0	Horizontal	Frankfurt	MK 77	Klima and Van Bree (1985)
115.0	48.0	Transverse	Frankfurt	K 4	Present report
Pilot whale, *Globicephala melas*					
123.0	-	Transverse	-	-	Schreiber (1916)

Table 1 continue

Common porpoise, *Phocoena phocoena*					
25.0	10.0	Transverse	Frankfurt	MK 74	Klima and Van Bree (1985)
28.0	11.0	Sagittal	Frankfurt	MK 71	Klima and Van Bree (1985)
39.0	14.0	Transverse	Frankfurt	MK 72	Klima and Van Bree (1985)
41.0	19.0	Transverse	Frankfurt	MK 73	Klima and Van Bree (1985)
42.0	18.0	Sagittal	Frankfurt	MK 70	Klima and Van Bree (1985)
51.0	22.0	Sagittal	Frankfurt	MK 67	Klima and Van Bree (1985)
60.0	24.0	Sagittal	Frankfurt	MK 64	Klima and Van Bree (1985)
69.0	27.0	Sagittal	Frankfurt	MK 75	Klima and Van Bree (1985)
70.0	29.0	Sagittal	Frankfurt	MK 62	Klima and Van Bree (1985)
95.0	45.0	Sagittal	Frankfurt	MK 61	Klima and Van Bree (1985)
107.0	46.0	Transverse	Frankfurt	MK 69B	Klima and Van Bree (1985)
125.0	55.0	Transverse	Frankfurt	MK 68B	Klima and Van Bree (1985)
134.0	60.0	Horizontal	Frankfurt	MK 76	Klima and Van Bree (1985)
167.0	-	Transverse	Frankfurt	MK 48	Klima and Van Bree (1985)
Narwhal, *Monodon monoceros*					
137	-	Horizontal	Utrecht	-	Klima and Van Bree (1985)
Sperm whale, *Physeter catodon*					
19.0	10.0	Sagittal	Frankfurt	MK 58	Klima et al. (1986)
37.0	14.0	Sagittal	Frankfurt	MK 51	Klima et al. (1986)
60.0	-	Dissection	Frankfurt	PH D2	Klima et al. (1986)
75.0	-	Transverse	Frankfurt	PC	Klima et al. (1986)
89.0	41.0	Sagittal	Frankfurt	MK 52	Klima et al. (1986)
97.0	45.0	Transverse	Frankfurt	MK 62 A	Klima et al. (1986)
102.0	47.0	Horizontal	Frankfurt	MK 66	Klima et al. (1986)
150.0	66.0	Transverse	Frankfurt	MK 63	Klima et al. (1986)
170.0	73.0	Sagittal	Frankfurt	MK 53	Klima et al. (1986)
315.0	145.0	Dissection	Frankfurt	PH 4	Klima et al. (1986)
385.0	180.0	Ct and nmr	Frankfurt	PH 6	Klima et al. (1986)
435.0	210.0	Dissection	Frankfurt	PH 3	Klima et al. (1986)
455.0	240.0	Dissection	Frankfurt	PH 2	Klima et al. (1986)
565.0	260.0	Dissection	Frankfurt	PH 1	Klima et al. (1986)
900.0	-	Dissection	Frankfurt	-	Klima et al. (1986)
Fin whale, *Balaenoptera physalus*					
156.0	-	Transverse	Frankfurt	MK 65	Present report

Table 1 continue

Minke whale, *Balaenoptera acutorostrata*					
455.0	-	Dissection	Frankfurt	3119	Present report
Humpback whale, *Megaptera novaeangliae*					
45.0	21.0	Transverse	Tokyo	-	Present report
90.0	37.0	Transverse	Tokyo	-	Present report
107.0	53.0	Sagittal	Utrecht	MK 54	Present report

macroscopical dissection. Several specimens were selected for modeling or for graphic reconstruction of the nasal structures, as described in the papers by Klima and Van Bree (1985), Klima et al. (1986) and Klima (1987a).

3 Synopsis of Structures of the Embryonic Nasal Skull

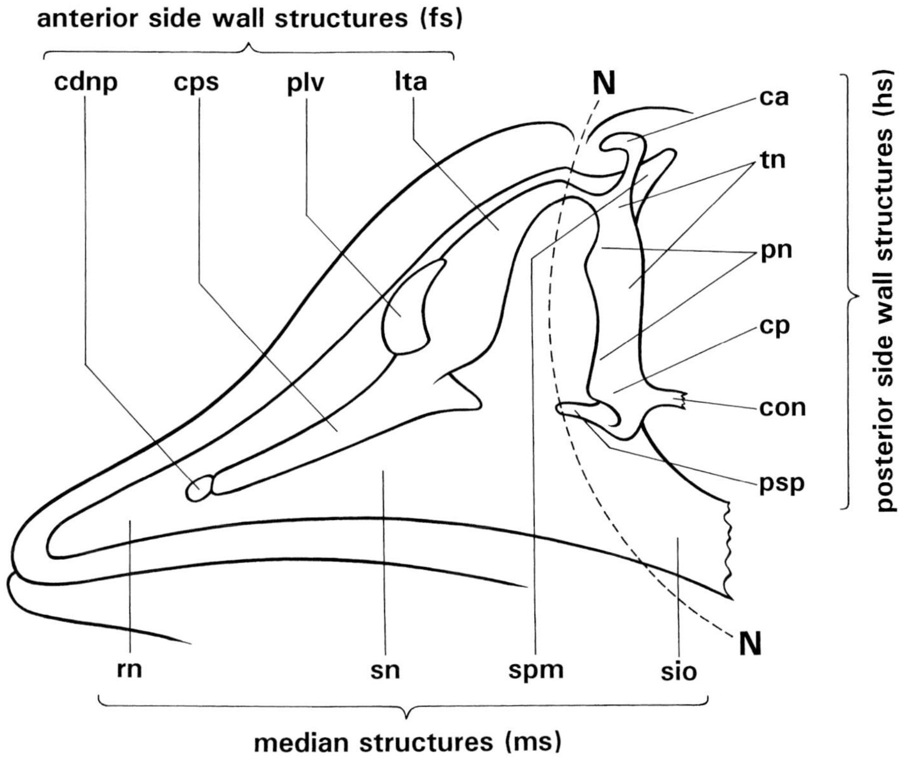

Fig. 2. Synopsis of the cartilaginous structures of the embryonic nasal skull of Cetacea in lateral view

The embryonic nasal skull of cetaceans is very much different from the scheme common to most other mammals. This is true not only of the presence or absence of the individual nasal structures but also of their number, shape and position, as well as of their later morphological and functional participation in the structure of the whole skull. In order to facilitate orientation in these unusual structures, a synopsis of the most important nasal structures is presented here as an introduction (Fig. 2), however, without giving any particulars. The division and terminology of the structures presented here will be consequentially adhered to in the subsequent chapters.

3.1 Cartilaginous Structures

The elements of the cartilaginous nasal capsule can be divided into three groups: (1) the median structures, (2) the posterior side wall structures, and (3) the anterior side wall structures.

The median structures (ms) form the nasal septum, or septum nasi (sn), in which the following structures can be discerned in the anterior-posterior direction: the rostrum nasi (rn), the septum nasi (sn), the spina mesethmoidalis (spm), and the septum interorbitale (sio).

The posterior side wall structures (hs) lie behind the nasal passage (N) and contain fused material of the original elements of the nasal roof, the tectum nasi, and the nasal wall, the paries nasi. In the anterior-posterior direction, or from top to bottom, they include the following structures : the cupula nasi anterior (ca), the tectum nasi (tn), the paries nasi (pn), the commissura orbitonasalis (con), the cupula nasi posterior (cp) and the processus paraseptalis posterior (psp). On the inside of the posterior side wall structures (Fig. 3a), it is possible to discern, in addition, the lamina cribrosa (lcr), the maxilloturbinale (mtx) and occasionally some other structures not mentioned here.

The anterior side wall structures (fs) lie in front of the nasal passage and comprise, on the one hand, isolated and on the other, fused elements of the original nasal floor, the solum nasi. In the anterior-posterior direction, they include the following structures: the cartilago ductus nasopalatini (cdnp), the cartilago paraseptalis (cps), the processus lateralis ventralis (plv) and the lamina transversalis anterior (lta).

3.2 Dermal Bones

In the formation of the nasal skull, three large dermal bones play a decisive part. All of them show a close connection with certain cartilaginous structures (Fig. 3b,c).

The praemaxillare, or premaxillary bone (pmx) accompanies the rostrum nasi and the anterior side wall structures. It participates in the upper anterior and lateral delimitation of the nasal passage.

The maxillare, or maxillary bone (mx), the largest dermal element of the nasal skull, is laterally adjacent to the praemaxillare and the vomer. In its posterior part, it is intimate with the posterior side wall structures. It participates in the medial anterior and lateral delimitation of the nasal passage.

The vomer, or vomer bone (vom), U-shaped in cross section, and longitudinally covering the whole ventral margin of the cartilaginous median structures, supports the cartilago paraseptalis and the processus paraseptalis posterior from ventral.

Among the smaller dermal bones, especially the nasale, or nasal bone (ns) should be mentioned, joining the posterior side wall structures dorsally and forming the upper posterior margin of the nasal passage. The palatinum and the pterygoid, not shown here, form the lower margin of the nasal passage.

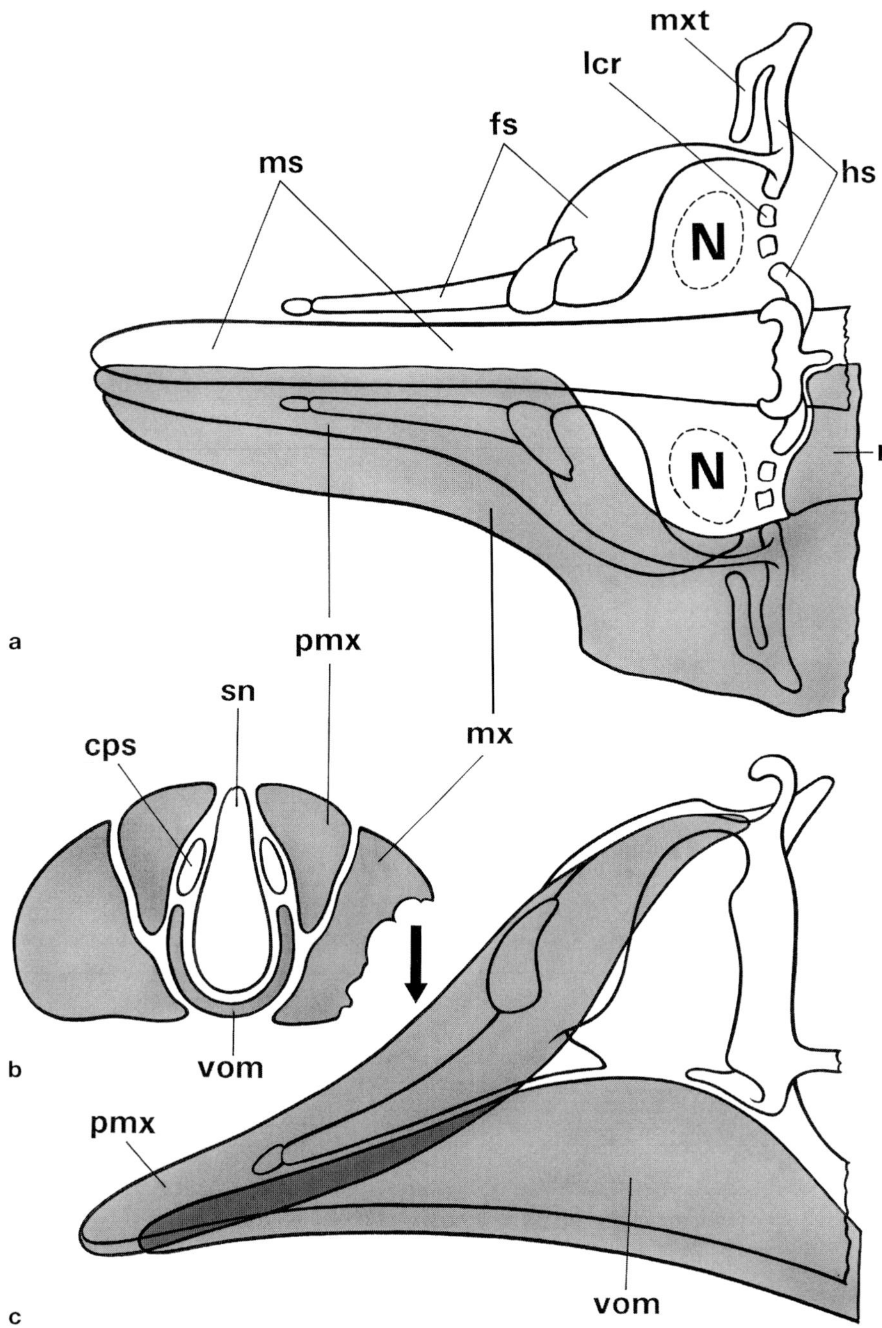

Fig. 3a-c. Synopsis of the cartilaginous structures and accompanying dermal bones of the embryonic nasal skull of Cetacea. Depending on the viewing direction, some of the elements have been omitted. **a** Horizontal projection. **b** Cross section; arrow indicates section plane of cross section in bottom drawing. **c** Lateral projection

4 Morphogeny of the Nasal Skull

4.1 Odontoceti

4.1.1 Spotted Dolphin *Stenella attenuata*

The most extensive material available for this study was that of *Stenella attenuata*, comprising an almost continuous developmental series of a total of 14 embryos ranging between 21 and 225 mm in size. Four of the embryos, 39, 43, 57 and 85 mm in size, have been partly studied in an earlier paper (Klima 1987a), the rest are presented here for the first time. The results obtained here from *Stenella* serve as a comparative basis for observations obtained from all other species. For this reason, they are discussed here in considerable depth.

It has appeared useful to divide the embryos examined into five groups according to their developmental stages.

Group I, Embryos 21, 22, 25 mm

The youngest embryos examined show in the region of the nasal skull, no more than an indistinctly defined mesenchyme condensation. Its individual parts cannot be precisely identified with the future cartilaginous structures of the nasal capsule. Also, there are no anlagen of dermal bones in the region of the nasal skull.

Group II, Embryos 39, 43.5, 44.5 mm

In these embryos a simple anlage of the nasal capsule can be discerned (Fig. 4a,b). The median structures claim the largest part of this anlage. Of these structures, the posterior inferior elements, i.e., the septum interorbitale and the basis of the septum nasi show the most advanced degree of development, consisting of young cartilaginous tissue. In contrast to that, the anterior part of the septum nasi, as well as its upper edge up to the anlage of the spina mesethmoidalis reaching far upwards, are not yet as advanced. With respect to histogenesis, these elements are still at the mesenchymatous developmental level and are not yet sharply defined from the surrounding tissue. This is in agreement with the general rule, according to which the histogenesis of the

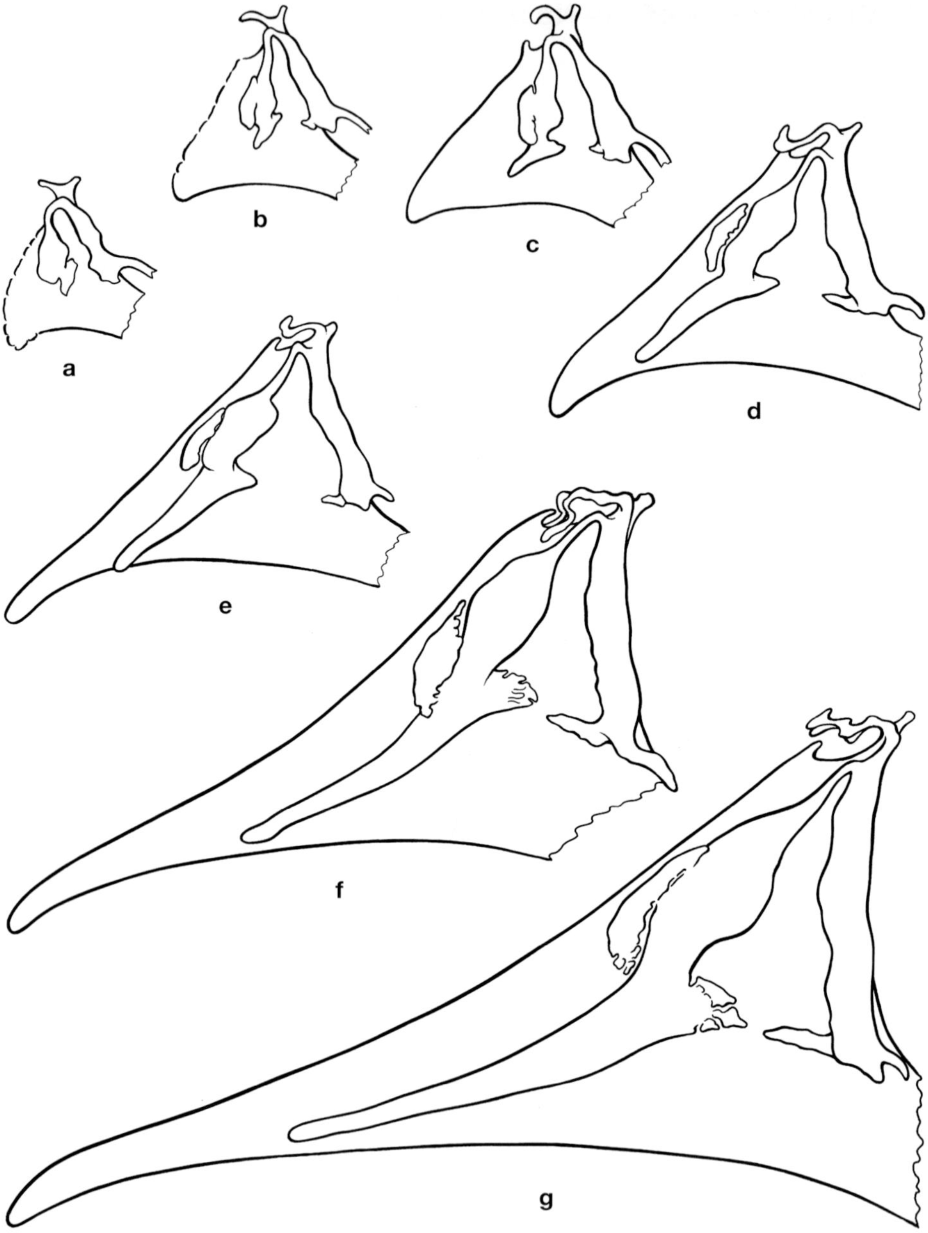

Fig. 4a-g. *Stenella attenuata*. Lateral views of the cartilaginous structures of the embryonic nasal skull. Size of the embryos in mm TL: **a** 39; **b** 44.5; **c** 57; **d** 71; **e** 76; **f** 153; **g** 225

chondrocranium proceeds from occipital towards rostral. A rostrum nasi is still not discernible. In this stage the whole nasal septum is approximately an equilateral triangle.

The posterior superior edge of the septum nasi below the spina mesethmoidalis passes into the posterior side wall structures laterally on both sides. Its material consists of precartilaginous mesenchyme. Its blurred contours permit a rather vague distinction between the following elements: a cupula nasi anterior, reaching fairly far upwards, a porous and discontinuous tectum nasi with lamina cribrosa, and a plate of the paries nasi protruding laterally. Below and posteriorly the paries nasi passes into a narrow and long commissura orbitonasalis communicating with ala orbitalis. The lower edge of the paries nasi is slightly rolled, this part obviously belonging to the cupula nasi posterior and containing a discernible low spine of the processus paraseptalis posterior protruding rostrally. The lateral edge of the paries nasi bears a low, inwardly turned wrinkle, invisible from the outside, the apparent anlage of the maxilloturbinale.

As a whole, the posterior side wall structures form a flat plate which incompletely surrounds the nasal cavity from above and posteriorly. This plate laterally accompanying the upper posterior edge of the septum nasi, shows much the same inclination as this edge. It passes from below posterior obliquely towards upper anterior at an angle of about 45° in the 39 mm embryo, and at a somewhat steeper angle in those 43.5 and 44.5 mm in size.

The anterior superior part of the paries nasi passes into a narrow connecting shaft which bends down sharply to join with the anterior side wall structures. The latter consist, histologically, of precartilaginous mesenchyme with indistinct outer limits. The circumferences of these mesenchymal condensations permit the following structures to be discerned. Above, a larger structure representing a continuation of the connecting shaft to the paries nasi. This structure is undivided in the 39 mm embryo, whereas in those 43.5 and 44.5 mm in size, it differentiates into a posterior anlage of the lamina transversalis anterior, and into an anterior one of the processus lateralis ventralis. The lower part of the anterior side wall structures is present as a small mesenchyme condensation, the anlage of the cartilago paraseptalis, probably together with the integrated material of the cartilago ductus nasopalatini.

In the youngest embryo 39 mm in size, this anlage initially points largely caudally and ventrally. In this way, it approximately repeats its original position typical of the nasal capsule of terrestrial mammals. Later, in the embryos 43.5 and 44.5 mm in size, it shifts somewhat more ventrally, i.e., perpendicular to the lower edge of the septum nasi.

No dermal bones are present in the region of the cartilaginous nasal capsule in the examined embryos of this stage.

Group III, Embryos 57, 71 and 76 mm

In these embryos the precartilaginous mesenchyme matures to young embryonic cartilaginous tissue and the whole nasal capsule shows more distinct contours (Fig. 4c,d,e). Intensive growth processes set in, changing the proportions and form of the anlage. Especially the rostral parts of the median structures grow rapidly rostrally. In this way the whole anlage of the nasal capsule increases in length and a long rostrum nasi begins to reach out anteriorly. The posterior edge of the septum nasi, together with the posterior side wall structures, become steeper, now being inclined at an angle of about 70° anteriorly. The spina mesethmoidalis is bent towards the posterior.

Among the posterior side wall structures, the greatest share falls to the paries nasi and the tectum nasi. Together with the lamina cribrosa, these two structures have been pressed together at the posterior edge of the septum to form a common flat plate. These structures cannot be precisely delimited. Also, the individual structures, forming relief projections on the inner side of the concave plate, cannot be unequivocally determined. Judging from their size, I consider the largest projection of the upper side wall, rolled towards medial, to be the maxilloturbinale, and the smaller projections more closely adjacent to the lamina cribrosa, to be the ethmoturbinale. From the lower edge of the plate, a projection forming a short arch points rostral medially, becoming intimate with the septum without fusing with it. Most probably, this is the material of the lamina transversalis posterior with the processus paraseptalis posterior. In the opposite direction, i.e., caudal laterally, a narrow band of the commissura orbitonasalis protrudes from the lower edge of the posterior side wall structures. In the 57 mm embryo, this projection is already connected on either side with the ala orbitalis. In the 71 mm embryo, it is asymmetric and connects the nasal skull with the ala orbitalis on the left side of the head only. On the right side it is only present as a blind projection. In the 76 mm embryo, it is present as a blind projection bilaterally.

Among the anterior side wall structures, three elements can be distinguished. Above, the lamina transversalis anterior, which is connected with the upper region of the paries nasi and thus also with the posterior side wall structures through a narrow bridge. In this section the lamina transversalis anterior is rather far away from the median plane, as it forms there the lateral wall of the nasal passage. Anterior and ventrally, it approaches, in a wide arch, the median plane and becomes intimate with the septum nasi without fusing with it. This lower section of the lamina transversalis anterior is especially massive. Its anterior edge comprises the material of the processus lateralis ventralis. In the 57 mm embryo, these structures are still united; while in those 71 and 76 mm, the processus lateralis ventralis is already separated and forms an independent element. The cartilago paraseptalis joins the lower edge of the lamina transversalis anterior. In the 57 mm embryo it is still small; in those 71 and 76 mm it has considerably increased in size, forming a long and heavy bar growing parallel with the rostrum nasi anteriorly, and together with it, accounting for the formation of the protruding nasal tip. The anterior tip of the cartilago paraseptalis probably contains material of the cartilago ductus nasopalatini. Posteriorly, the cartilago paraseptalis emits a short projection, the processus paraseptalis anterior, the tip of which points towards the processus paraseptalis posterior of the posterior side wall structures.

Besides the cartilaginous structures, this stage already shows the first anlagen of dermal bones, which participate in the formation of the nasal skeleton. In the smallest 57 mm embryo the anlagen are still indistinct and hard to define. In the 71 mm embryo and especially in the biggest one of 76 mm, their contours are already distinct. At the same time, there appears a conspicuous connection of the praemaxillare with the lamina transversalis anterior, and of the vomer with the cartilago paraseptalis. The septum nasi and rostrum nasi are accompanied, along their inferior edge, by the vomer and laterally by the praemaxillare and maxillare. In the posterior, i.e., occipital region, the young bone tissue shows a higher degree of developmental progression than in the anterior, i.e., rostral region. The foremost parts of the rostrum nasi and the cartilago paraseptalis are still not accompanied by any dermal bones.

Group IV, Embryos 85, 89, 115 mm

The embryos of this size group occupy a key position in the present investigations. Here the morphogeny of the nasal skull reaches its peak. All of the most important embryonic nasal structures are fully developed and assume their approximate final position (Figs. 5-8). The cartilaginous elements of the endocranium show a distinct connection with the dermal bones of the exocranium. The whole nasal skull has reached an ontogenetic degree which is particularly suitable for direct comparison with other species and which also permits comparison with the nasal skeleton of adults.

The median structures, consisting of young cartilaginous tissue, form in their entirety, an irregular triangle projecting far anteriorly in a long rostrum. Its medial part is formed by the high septum nasi. Its superior edge is slightly concave posteriorly in the region of the nasal ducts. Still farther occipitally, the upper part of the septum nasi connects with the posterior side wall structures, ending behind them in a short projection of the spina mesethmoidalis. The posterior, i.e., occipital edge, of the septum nasi falls down perpendicular ventrally, forming an angle of about 90° with the longitudinal axis of the whole anlage. Posteriorly and ventrally the septum nasi passes into the septum interorbitale, the most massive part of the whole nasal capsule. The lower parts of the septum interorbitale, the septum nasi and the rostrum form a continuum which, like a strut passing in the longitudinal direction, constitutes a kind of a firm base of the nasal capsule. This base is slightly concave on its lower side and tapers off in the direction from caudal towards rostral.

The posterior side wall structures as a whole are flattened paired plates which, like powerful wings, spread perpendicularly from the caudal edge of the septum nasi laterally on both sides. They accompany the caudal edge of the septum nasi all along their length up to the beginning of the septum interorbitale. Also, they lie at the same level as the occipital edge of the septum nasi, i.e., they fall off abruptly, forming an angle of about 90° with the longitudinal axis of the nasal capsule basis. The medial part of the posterior side wall structures form the posterior wall of the nasal passage. The lateral part becomes a component of the skull wall between the anterior viscerocranium and the posterior neurocranium.

Among the posterior side wall structures the following particular elements can be distinguished. Above is a rounded cupula nasi anterior tapering towards the anterior into a narrow projection, broken laterally, forming the rim of the nasal opening. Below that, the tectum nasi which connects above medially with the posterior edge of the septum nasi but is distant from it below. Ventrally the tectum nasi passes continually into the lamina cribrosa and the paries nasi.

The lamina cribrosa occupies the medial part of the anlage. It is also separated from the posterior edge of the septum nasi by a gap. In its flat wall there already are, on either side, three irregularly shaped, small apertures permitting entrance to filaments of the nervus olfactorius, which is heavily reduced, however.

The paries nasi occupies the lateral part of the anlage. It forms a simple, flat ledge, the lateral edge of which is turned at a sharp angle back rostral medially. This fold, accompanying the nasal passage all along its length laterally, corresponds to the

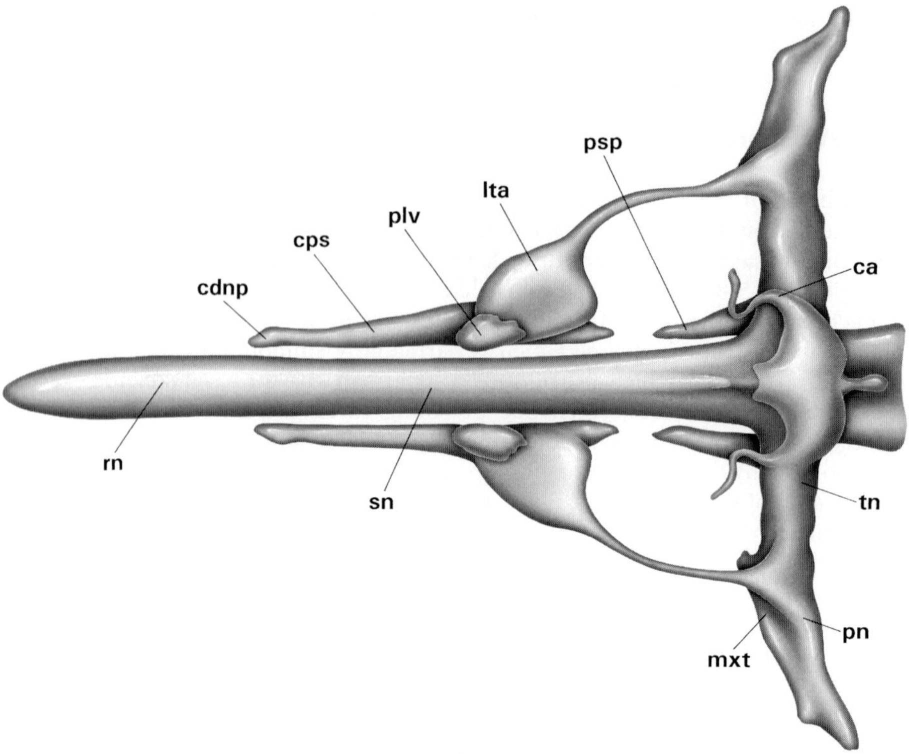

Fig. 5. *Stenella attenuata* embryo, 89 mm TL. Dorsal view of the cartilaginous structures of the embryonic nasal skull

maxilloturbinale. Its form and position become apparent only in the cross section through the nasal capsule, and as the cupula nasi posterior, it forms the incomplete posterior, more precisely, the lower rim of the nasal passage. From this rolled part arises, rostrally, a narrow projection, the processus paraseptalis posterior. A connection between the paries nasi and the ala orbitalis in the form of a commissura orbitonasalis is no longer present. The commissura orbitonasalis has been reduced, on either side, to a short, blind projection.

The upper edge of the paries nasi sends, rostrally, a narrow connecting shaft to the lamina transversalis anterior, a part of the anterior side structures. This connecting shaft forms another through arch around the upper part of the nasal passage, immediately beneath the lateral projection of the cupula nasi anterior. In front, the connecting shaft bends medially, continuing in the massive anlage of the lamina transversalis anterior. The latter forms the anterior terminal part of the whole arch. With its medial surface, the lamina transversalis anterior approaches the lateral surface of the septum nasi but does not fuse with it. A narrow gap is preserved between these two structures. After all, this is true of all the elements of the anterior side wall structures which, while intimate with the median structures, never fuse with them in any place.

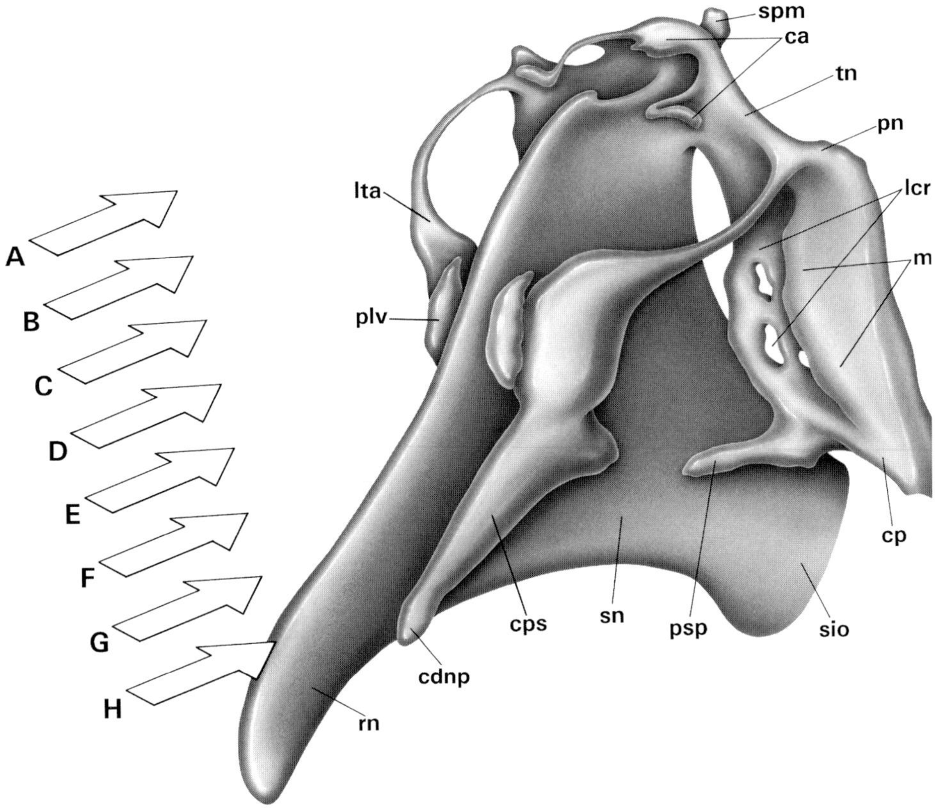

Fig. 6. *Stenella attenuata* embryo, 89 mm TL. Cartilaginous structures of the embryonic nasal skull in an oblique view from frontal and lateral direction. The *arrows* indicate the section planes in Fig. 7

From the anterior side of the lamina transversalis anterior sprouts a small, somewhat irregularly shaped anlage of the processus lateralis ventralis. The lower surface of the lamina transversalis anterior continues into the cartilago paraseptalis. The latter bears posteriorly a short projection, the processus paraseptalis anterior. The main part of the cartilago paraseptalis extends far rostrally in the form of an irregular, laterally flattened triangle, the upper edge of which points in about the same direction as the parallel upper edge of the rostrum nasi. The anterior part of the cartilago paraseptalis is slightly curved downwards and its tip ends about the middle of the rostrum nasi on its lower edge. The tip of the cartilago paraseptalis bears a slight protrusion, the probable material for the cartilago ductus nasopalatini.

In all embryos of this size the anlagen of dermal bones are fully developed. They become intimate, from the outside, with the cartilaginous structures, forming together the characteristic structure of the embryonic nasal skull.

The praemaxillare accompanies, from the lateral direction, the upper edge of the rostrum nasi and the anterior side wall structures all along their length. Its anterior tip

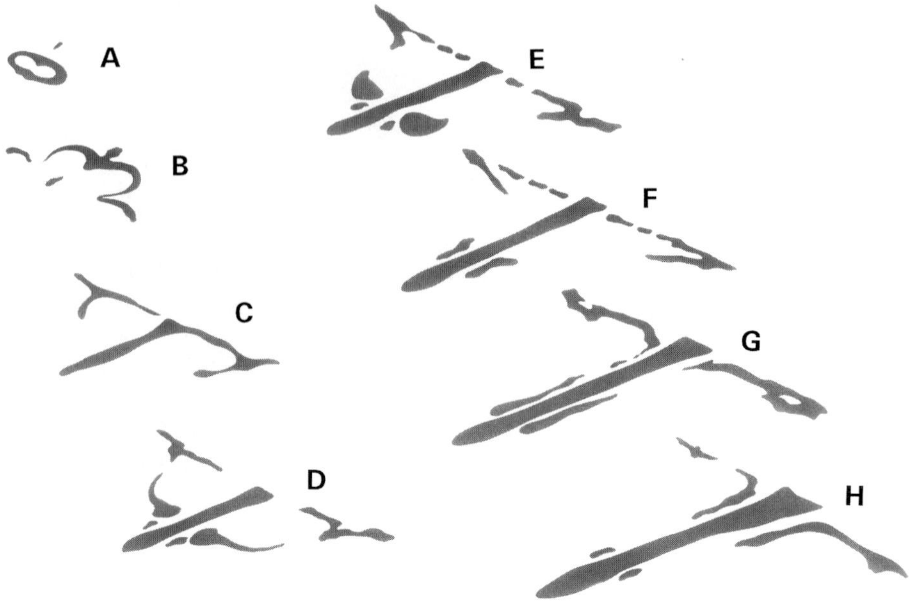

Fig. 7A-H. *Stenella attenuata* embryo, 89 mm TL. Cartilaginous structures of the embryonic nasal skull, horizontal sections. **A-H** Section planes corresponding to the *arrows* in Fig. 6

does not quite attain the tip of the cartilaginous rostrum nasi which overreaches it slightly, and which is conspicuous particularly in the smaller embryos of this group. The posterior tip of the praemaxillare ends behind the lamina transversalis anterior, lateral to the connecting shaft with the paries nasi.

The maxillare accompanies, from the lateral direction, the praemaxillare all along its length, reaching posteriorly behind the posterior side wall structures up to the other dermal bones, the nasale and the frontale, which participate in the formation of the anterior wall of the neurocranium. The anterior tip of the maxillare is situated at about the same level as the anterior tip of the praemaxillare; that is to say, even the maxillare is slightly overreached by the cartilaginous rostrum nasi. In its lower region the maxillare is intimate with the vomer. The vomer forms a cover of the lower edge of the rostrum nasi and the septum nasi, passing up to the septum interorbitale. On cross section, the vomer is typically U-shaped. Its anterior tip does not reach as far rostrally as those of the praemaxillare and maxillare. Its upper edge accompanies the cartilago paraseptalis all along its length up to behind the processus paraseptalis anterior, continuing as far as below the processus paraseptalis posterior of the posterior side wall structures.

Group V, Embryos 153, 225 mm

In the preceding Group IV, the morphogenetic processes of the nasal skull have been largely completed. The subsequent development is chiefly determined by intense growth processes. Therefore, no fundamental changes can be expected to occur in the

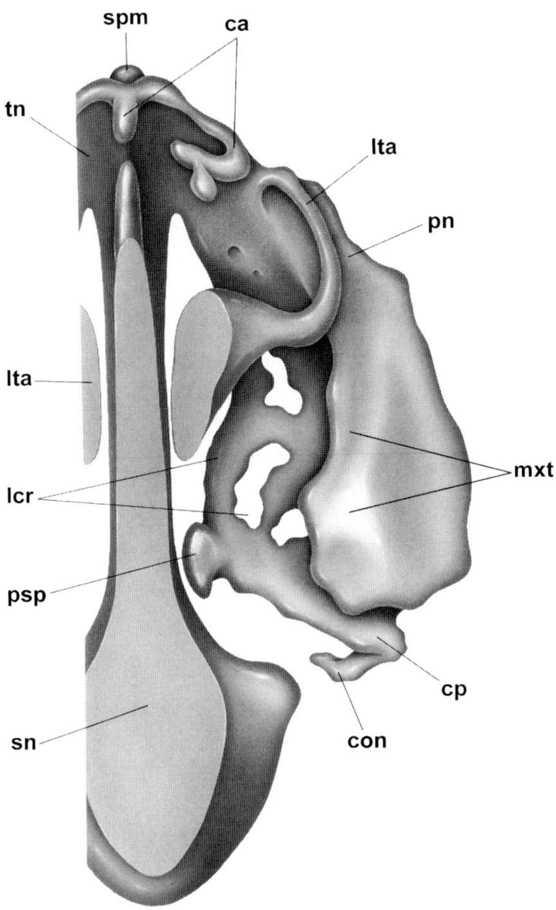

Fig. 8. *Stenella attenuata* embryo, 115 mm TL. Frontal view of the cartilaginous structures of the embryonic nasal skull and left half of the posterior side wall structures. The anterior part of the skull reconstruction has been removed by a cross section through septum nasi (*sn*) and lamina transversalis anterior (*lta*)

biggest embryos, 153 and 225 mm in size, comprising Group V. The differences pertain, first of all, to the developmental state of the tissues as well as to the form and proportions of the whole anlage. But for that, all the elements of the nasal skull, i.e., the cartilaginous structures as well as the dermal bones, show much the same developmental degrees as those described in the embryos of Group IV (Figs. 4f,g, 9).

Histologically, the cartilaginous structures consist of a typical embryonic hyaline cartilage. The cell mass no longer exceeds that of the extracellular substance, as was the case with the smaller embryos . The amount of the two constituents is largely equal. The large number of dividing cells indicates that this is an intensely growing, young cartilaginous tissue. It is also indicated by the heavy vascularization of the tissue, a condition not found in the earlier stages. The blood vessel channels form considerably branched systems that penetrate from the perichondrium deep into the tissue layers,

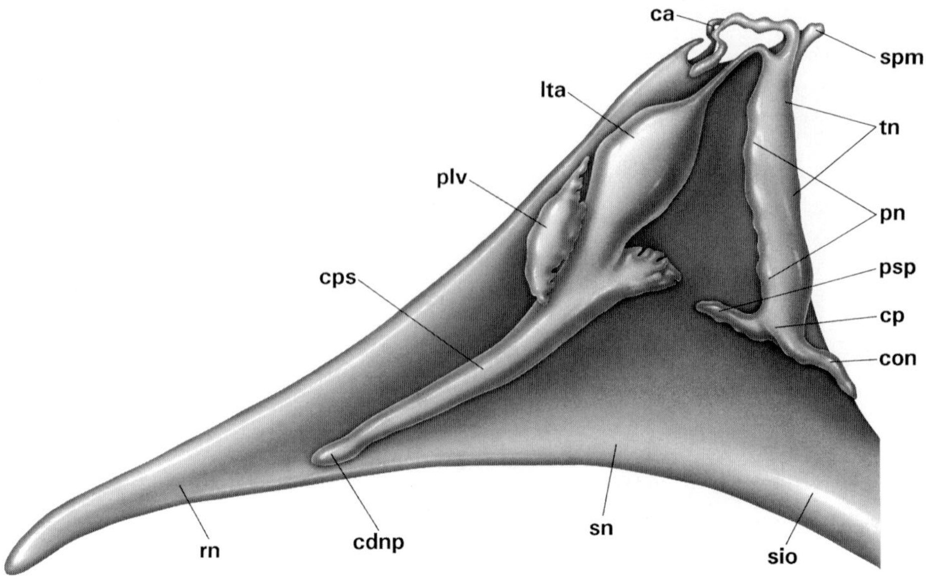

Fig. 9. *Stenella attenuata* embryo, 153 mm TL. Lateral view of the cartilaginous structures of the embryonic nasal skull

thus providing the growing tissues with a rich blood supply. The heaviest vascularization is found in the massive parts of the cartilaginous nasal capsule, in particular the septum interorbitale, the septum nasi, the rostrum nasi and the lamina transversalis anterior. All cartilaginous structures possess, on their surface, a distinct layer of the perichondrium, quite distinctly defined from the surrounding tissue.

Even the histological processes in the dermal bones have advanced considerably. In the smaller embryos the outlines of these structures were often indistinct, as there was no outer continuous layer of periost and the inside consisted of a confused network of cells and streaming basal substance. In the embryos of this size group, the desmal ossification of the dermal bones is well under way. The external surface of these structures is covered with an almost complete periost, defining them distinctly from the surroundings and indicating the precise form of the anlagen. The osteoid, the osseous basic substance, is deposited inside the anlagen in the form of massive trabeculae. Intense growth processes take place.

As far as the changes in proportions and form are concerned, most of them are visible in the rostral region of the anlage. The cartilaginous rostrum nasi becomes protracted into a narrow bar reaching far towards the anterior region, thereby considerably changing the proportions. While the length: height ratio was around 1:1 in Group I, in the embryos of Group V it is about 2:1. Of the cartilaginous structures, not only the rostrum nasi proper, but also the cartilago paraseptalis take a decisive part in this prolongation. Also, all dermal bones in this rostral region, i.e., the praemaxillare, maxillare and vomer, grow far towards the anterior region like narrow accompanying structures. Their anterior tips attain the same level as the anterior tip of the cartilaginous rostrum nasi. This also terminates the still existing advance in the growth of the cartilaginous anlage over that of the desmal ones, and the entire complex of the rostrum forms a morphological entirety. At the same time, from now on, the nasal structures are out of consideration, as the complex rostrum has nothing to do with the nasal skeleton but belongs to the jaw apparatus.

In other regions of the embryonic anlage of the nasal skull no such consequential changes take place. The posterior projection of the cartilago paraseptalis, i.e., the processus paraseptalis anterior, becomes longer and broader, its surface bearing several irregular tissue protrusions. With its tip, the processus paraseptalis anterior approaches that of the processus paraseptalis posterior which, at this stage, reaches comparatively far anteriorly from the posterior side wall structures. The two projections form together a wide open arch above the edge of the vomer.

The inclination of the posterior edge of the septum nasi, the same as that of the posterior side wall structures, is still somewhat steeper, as in Group IV. It is inclined caudally at an angle of 90° or even more, measured from the longitudinal axis of the whole anlage. In the biggest embryo examined, 225 mm in size, these structures begin to bend slightly. They are slightly concave from the caudal, thereby fitting with the rounded form of the neurocranium lying behind.

4.1.2
Common Dolphin *Delphinus delphis*

In all, nine embryos of this species, 24 to 131 mm in size, were available (Figs. 10,11). By the degree of their development, they can be classified in the same five size groups as in *Stenella*. All major morphogenetic processes show much the same course as in *Stenella*. This pertains to the maturation of tissues as well as to changes in the shape and proportions of the whole nasal skull, the same as the relationships between the cartilaginous structures and the dermal bones.

The biggest embryonic *Delphinus* examined, 131 mm in length (Fig. 11), is most suitable for a detailed description of the nasal skull, and for comparison with other species. In this embryo the morphogeny of the nasal skull is largely completed, all nasal structures are present in their distinct forms, and intense growth processes have begun. For the sake of comparison, this embryo can be classified in Group V, and it can be directly compared with the embryonic *Stenella* 153 mm in size, also included in Group V.

The embryonic nasal skull of *Delphinus* possesses the same cartilaginous structures as that of *Stenella*, in an almost identical form and position. The two genera may hardly be unequivocally identified on the sole basis of their embryonic nasal capsules.

The median structures consist of a long and narrow rostrum nasi, passing into a high septum nasi more caudally. The upper edge of the septum nasi is slightly concave in the region of the cupula nasi anterior and it tapers caudally into a pointed spina mesethmoidalis. The latter is somewhat longer than in *Stenella*. Thereafter, the posterior edge of the septum nasi falls off abruptly downwards at an angle of about 90° with the longitudinal axis of the whole nasal capsule. Below, the base of the septum nasi continues into a robust septum interorbitale towards the caudal direction.

The posterior side wall structures are composed as follows. At the top, there is the cupula nasi anterior, consisting of somewhat irregularly shaped cartilaginous bands,

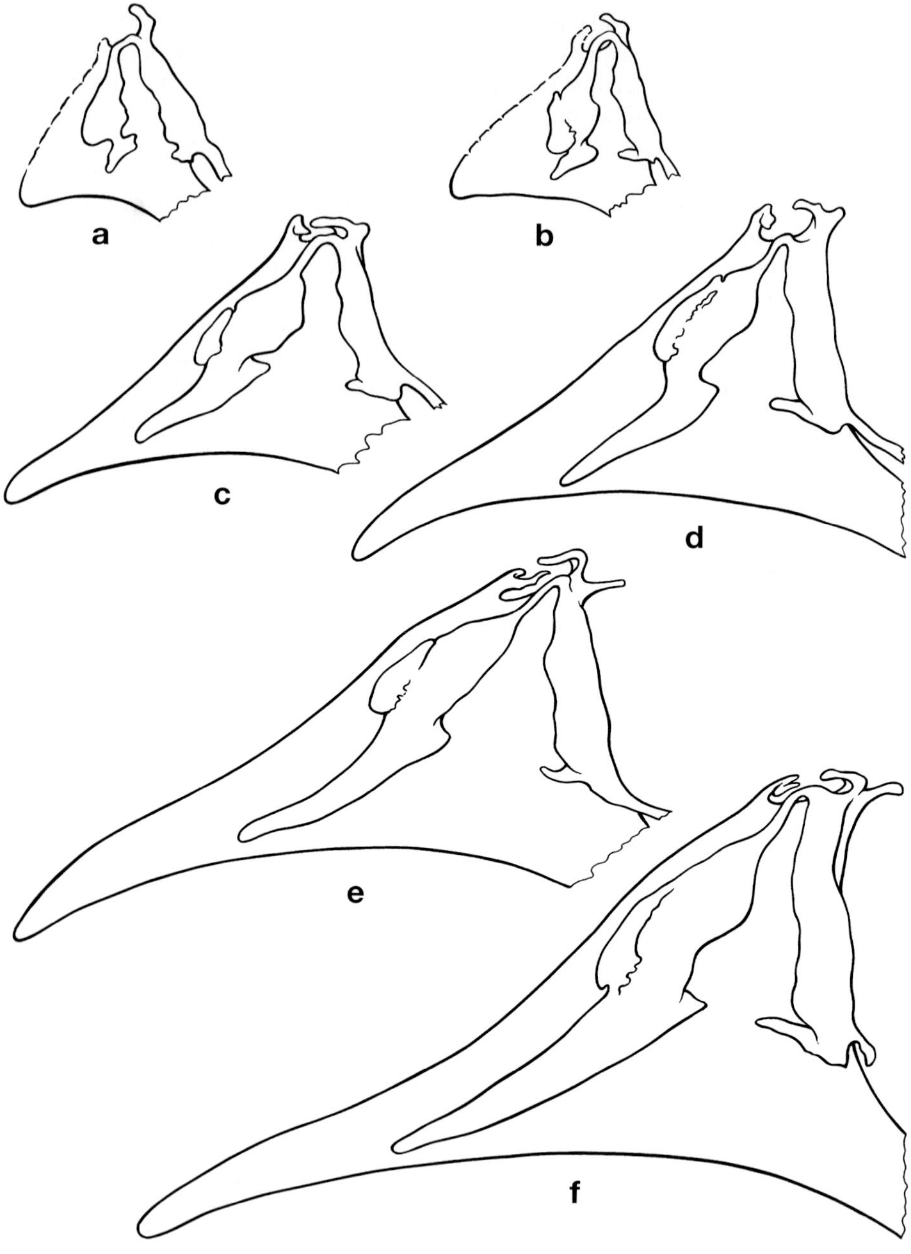

Fig. 10a-f. *Delphinus delphis*. Lateral views of the cartilaginous structures of the embryonic nasal skull. Size of the embryos in mm TL: **a** 42; **b** 46; **c** 82; **d** 101; **e** 113.5; **f** 131

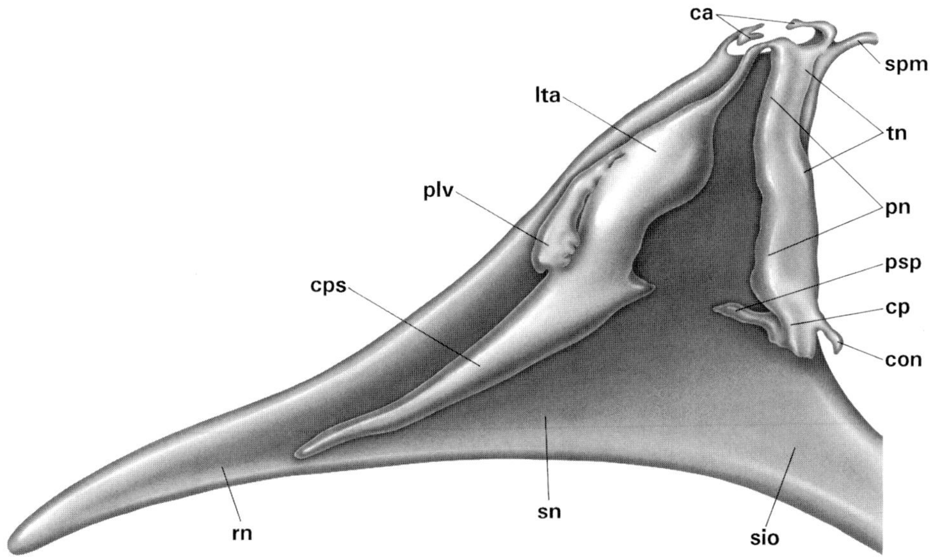

Fig. 11. *Delphinus delphis* embryo, 131 mm TL. Lateral view of the cartilaginous structures of the embryonic nasal skull

the form of which differs from that in *Stenella*. Besides, tissue protrusions of the upper edge of the septum nasi participate in the cartilaginous rim of the nasal opening, which is not the case in *Stenella*. Below the cupula nasi follows, in the median region, the tectum nasi, still forming a T-shaped connection with the septum nasi dorsally, but separating from it farther below and passing, without any visible boundary, into a perforated lamina cribrosa. The lateral region of the posterior side wall structures is the paries nasi, the edge of which recedes rostrally and to the inside in the form of a flat fold, the maxilloturbinale. The lowest region of the posterior side wall structures consists of a flattened fold of the cupula nasi posterior from which the processus paraseptalis posterior protrudes rostrally. The lower caudal region of the posterior side wall structures contains no more than a blind projection of the commissura orbitonasalis which, in the younger stages, still forms a connection with the ala orbitalis. The same situation is also found in *Stenella*.

The posterior side wall structures pass into the anterior ones through a narrow cartilaginous bridge in the region of the upper paries nasi. They consist of a massive lamina transversalis anterior that continues, towards the anterior region, in a semi-isolated processus lateralis ventralis and, ventrally, in a long band of the cartilago paraseptalis. The anterior tip of the cartilago paraseptalis shows no structure worth mentioning as a cartilago ductus nasopalatini. The posterior tip of the cartilago paraseptalis forms a somewhat smaller processus paraseptalis anterior than that found in *Stenella*.

The dermal bones show the same topographic relationships as those of *Stenella*. It is important that even in *Delphinus* the praemaxillare reaches beyond the lamina transversalis anterior, ending only at the bridge to the paries nasi.

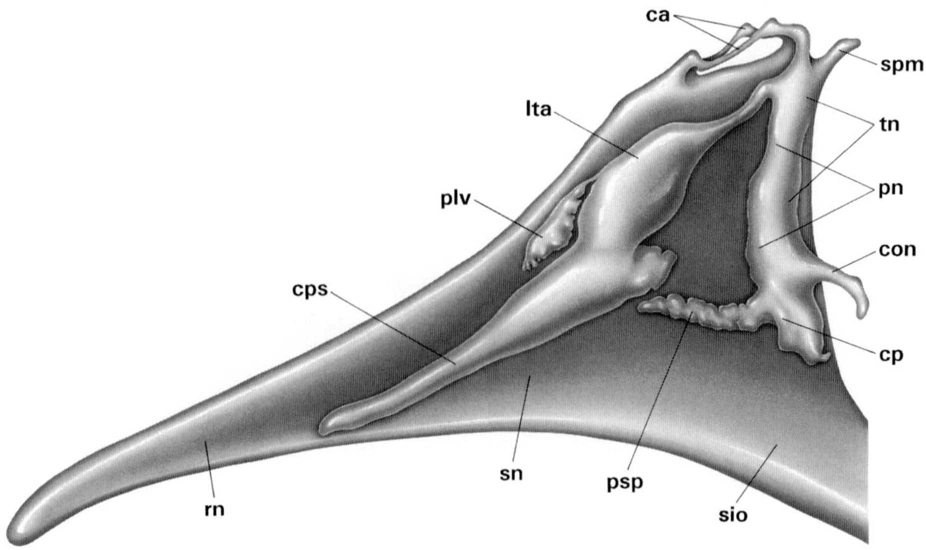

Fig. 12. *Lagenorhynchus albirostris* embryo, 115 mm TL. Lateral view of the cartilaginous structures of the embryonic nasal skull

4.1.3
White-beaked Dolphin *Lagenorhynchus albirostris*

In this species, two embryos, of 107 and 115 mm, were available for examination. We have already reported on the smaller one in an earlier paper (Klima and Van Bree 1985). In addition, De Burlet (1914b) examined an embryonic *Lagenorhynchus albirostris*, without having given its size, which I would estimate at about 90 mm. All three embryos are comparable with those of *Stenella* in Group IV by their size as well as the developmental stage of their nasal skulls.

The similarity to the nasal skull of Stenella is particularly conspicuous in the biggest embryo examined (Fig. 12). On the whole, the cartilaginous structures are present, the same as in *Stenella*. The only difference worth mentioning is that the processus paraseptalis posterior is very thick, reaching far anteriorly, as far as below the cartilago paraseptalis. The processus paraseptalis posterior attains a size larger than in all other cetacean species examined. Histologically, it consists of a porous cartilage, the surface of which bears numerous irregularities. A penetrating commissura orbitonasalis is present only in the smallest of the embryos, whereas in the two biggest ones it only forms a blind processus. In the cupula nasi anterior, the projections of the tectum nasi fuse with the narrow cartilaginous strips coming from the septum nasi to form a rim of the nasal opening. As far as the dermal bones are concerned, they show the same development and relationships to the individual cartilaginous structures as in *Stenella*.

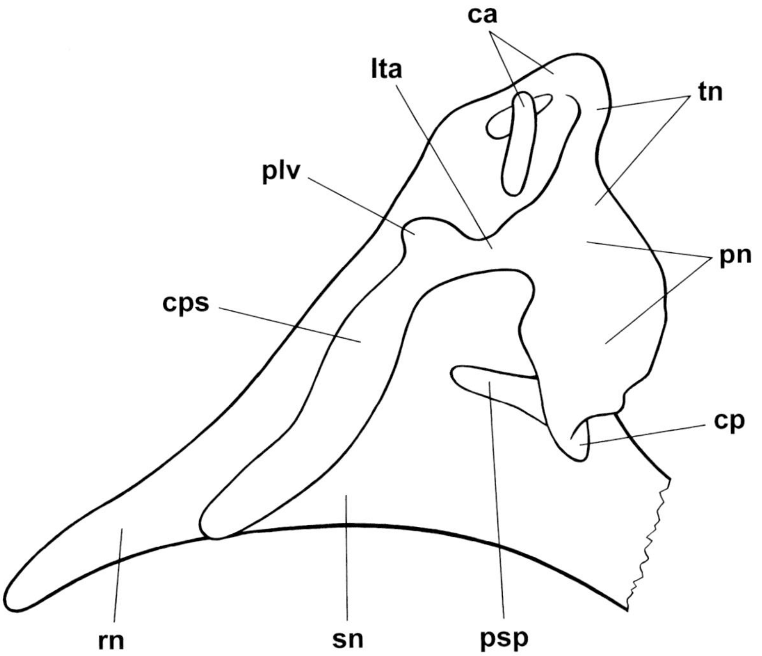

Fig. 13. *Globicephala melas* embryo, 123 mm TL. Lateral view of the cartilaginous structures of the embryonic nasal skull. (From Schreiber 1916)

4.1.4
Pilot Whale *Globicephala melas*

No material of this species has been available to me; nevertheless, the findings in an embryonic *Globicephala* 123 mm in size, published by Schreiber (1916) and later adopted by De Beer (1937) in his monograph, are illustrative to the extent that they permit comparison with other toothed whales. The embryonic *Globicephala*, 123 mm long, is comparable with the embryonic *Stenella* in Group IV.

The basic form of the cartilaginous nasal skull of *Globicephala* (Fig. 13), shows a high degree of similarity to that of *Stenella*. In particulars, however, several differences can be found. The cupula nasi anterior has no independent processus, but like that of Lagenorhynchus, it is largely integrated into the tectum and septum nasi. The posterior side wall structures are relatively broad and massive, the narrow processus paraseptalis posterior reaching rather far towards anterior. A massive cartilaginous bridge develops between the paries nasi from the posterior side wall structures and the lamina transversalis anterior from the anterior side wall structures. The anterior side wall structures have fused to form a single bar in which only slight elevations or incrassations indicate the probable position of the lamina transversalis anterior, the processus lateralis ventralis, the cartilago paraseptalis and the cartilago ductus nasopalatini.

4.1.5
Common Porpoise *Phocoena phocoena*

In all, 14 embryos of this species have been examined, their sizes ranging between 25 and 167 mm. A detailed report on them was presented in an earlier paper (Klima and Van Bree 1985). In addition, De Burlet (1913a) described two embryos of 48 and 92 mm. The whole material covers all five groups comparable with those of *Stenella.* It permits giving an almost uninterrupted picture of the morphogenetic processes of the nasal skull. These processes will be discussed later in this paper. At this time only the essential characteristics of the embryonic nasal skull are summarized.

Comparing embryos in Groups IV and V between *Phocoena* and *Stenella,* it becomes apparent that, following the nasal skull of *Physeter,* that of *Phocoena* shows the greatest difference from that of *Stenella* of all Odontoceti examined (Fig. 14). Its general form is lower, more regular and somewhat simplified.

The median structures form an almost equilateral triangle, the upper side of which, in contrast to almost all other Odontoceti, is not curved upwards but slightly downwards. In this way even the rostrum nasi gets the form of a triangle and not a kind of a bar. The posterior edge of the median structures falls off abruptly, so that even its upper tip points towards posterior. The spina mesethmoidalis is just moderately developed.

The posterior side wall structures are considerably massive and not as heavily flattened as in *Stenella.* The cupula nasi anterior forms a simple rounded structure. Its peripheral components are indistinctly or not at all connected with it, and form independent small cartilaginous parts around the nasal opening. In the interior relief of the posterior side wall structures there are, besides the maxilloturbinale, additional, distinctly visible convexities, probably corresponding to the ethmoturbinals. The cupula nasi posterior is relatively broad and its rounded structure reaches fairly far rostrally. It covers the short processus paraseptalis posterior. The commissura orbitonasalis is absent.

There is no connection between the posterior and anterior side wall structures like that found in all other Odontoceti examined, except in *Physeter.* There is a gap between the paries nasi and the lamina transversalis anterior where otherwise a cartilaginous bridge is found. The lamina transversalis anterior is large and heavily flattened from the lateral direction. It reaches rather far anteriorly, slightly protruding over the upper edge of the septum nasi. Its conspicuous anterior projection could contain material of the processus lateralis ventralis, but the two structures form a unit without any visible division. The cartilago paraseptalis, joining the lamina transversalis anterior from below, has the shape of a broad triangular plate. There is no independent cartilago ductus nasopalatini.

The dermal bones show the same connection with the cartilaginous structures as in all other Odontoceti examined. The posterior tip of the praemaxillare protrudes over the tip of the lamina transversalis anterior and ends at the anterior edge of the paries nasi, thereby replacing the missing cartilaginous bridge between the anterior and posterior side wall structures.

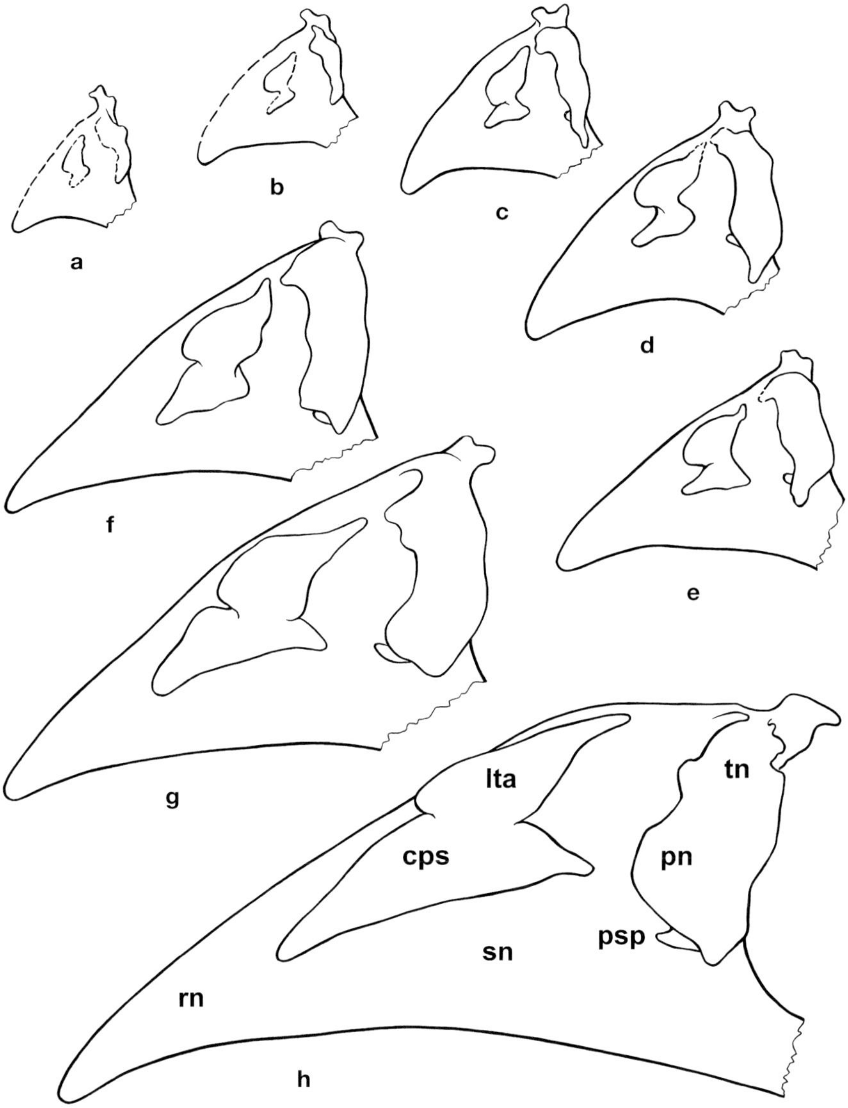

Fig. 14 a-h. *Phocoena phocoena*. Lateral views of the cartilaginous structures of the embryonic nasal skull. Size of the embryos in mm TL: **a** 41; **b** 51; **c** 60; **d** 69; **e** 70; **f** 95; g 134; **h** 167

4.1.6
Narwhal *Monodon monoceros*

A single embryo, 137 mm in size, was available, being comparable with those of *Stenella* in Group V. The chondrocranium of the same specimen had been examined by Eales (1950), however without any particulars being given on the ethmoidal region.

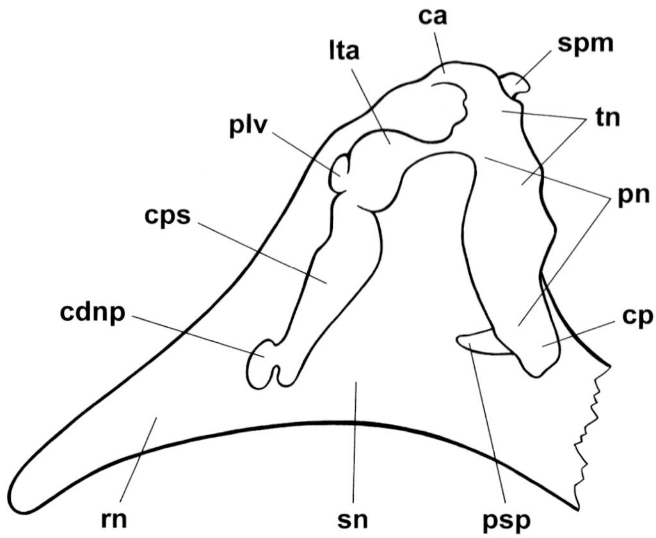

Fig. 15. *Monodon monoceros* embryo, 137 mm TL. Lateral view of the cartilaginous structures of the embryonic nasal skull

Therefore, we later made a reconstruction of the nasal capsule and carefully examined the whole nasal skull (Klima and Van Bree 1985).

Compared to that of *Stenella*, the cartilaginous nasal capsule of *Monodon* is somewhat more robust and not as slim (Fig. 15). The septum nasi is fairly high and also the rostrum nasi forms a perpendicular plate rather than a rounded bar. The posterior side wall structures show no appreciable differences from those of *Stenella*. The commissura orbitonasalis is absent. From the paries nasi, a rather thick bridge reaches to the massive lamina transversalis anterior from the anterior side wall structures. A small, semi-isolated processus is seen at its anterior edge, the processus lateralis ventralis. The lamina transversalis anterior continues obliquely towards the lower rostral region into the cartilago paraseptalis. The two structures are mutually separated by a deep incision but otherwise they form a coherent band. The cartilago paraseptalis is broad and short compared to the same structure in *Stenella*. Close to its rostral tip there is a small, semi-isolated cartilaginous part, the cartilago ductus nasopalatini, which is almost completely fused with the cartilago paraseptalis in *Stenella* as well as in all other Cetacean examined. In *Monodon* this structure has retained its independence, and a ductus nasopalatinus can even be demonstrated in its proximity (see Fig. 49).

As far as the dermal bones are connected, they show approximately the same development as in *Stenella* except for the vomer, which is distinctly shorter. Its anterior edge does not reach up to the tip of the rostrum nasi but ends under the cartilago ductus nasopalatini.

4.1.7 Sperm Whale *Physeter catodon*

In all, 15 embryos 19–900 mm in size were available. The material had already been used to examine the morphogeny of the nasal skull (Klima et al. 1986). It comprises all developmental stages corresponding to Groups I to V in *Stenella,* thus permitting an almost continuous study of the changes occurring during early ontogeny and their comparison with those occurring in other species. The changes will be discussed in detail in a different connection. At this time, the most important characteristics of the nasal skull are briefly summarized, as they can be observed in the embryos 150–315 mm in size, comparable by their developmental degree to those in Group V of *Stenella.*

The development of the external shape of the head of the sperm whale has been reported by Paliza (1964) and Nishiwaki et al. (1983), however without giving consideration to the connections with skull morphogeny. The most complete examinations in this line have been devoted to the development of the dermatocranium (Kuzmin 1969, 1974, 1976). Yet even those findings can only partly be compared with our results, as they pertain mainly to the later developmental stages and consider exclusively the dermal bones, without paying attention to the connections with the cartilaginous structures.

At first sight, the embryonic nasal skull of the sperm whale differs very strongly from that of the other toothed whales. In fact, however, even this unusual structure corresponds to the general basic model of the nasal capsule of all whales, although with differently proportioned individual elements (Figs. 16–18).

One of the most conspicuous traits is the fact that the anterior edge of the median structures, the septum nasi, produces two powerful processes. The inferior one, being a continuation of the inferior edge of the septum nasi, points anteriorly in almost horizontal position. This process corresponds to the rostrum nasi but is much longer and narrower than at the comparable stages of other toothed whales. The upper process rises from the anterior end of the tectum nasi. Together with the misshapen anlagen of the two cupulae nasales anteriores, the right one of which still remains underdeveloped, the material of the tectum nasi spreads obliquely anteriorly and dorsally. As a whole, these structures form a long cartilaginous bar running diagonally through the whole soft anterior part of the head, which I denote as the nasal roof cartilage.

But for the just mentioned singularity in the region of the anterior tectum nasi, the posterior side wall structures have basically the same structure as in all other toothed whales examined. Together with the posterior edge of the septum nasi, they fall off at a steep angle of about 75°.

The lamina cribrosa lies in the medial part of the posterior side wall structures. At first it is perforated, on either side, by just one opening, the fenestra cribrosa; later, i.e., from the stage around 150 mm onwards, by two openings, the fenestra cribrosa being divided into two openings by a cartilaginous bridge.

The lateral part of the complex of the posterior side wall structures corresponds largely to the paries nasi. The anterior, or more precisely, the upper part of the paries nasi inserts, over a small bridge, into the lamina transversalis anterior of the anterior side wall structures. This connection is absent from the left side. Thereby the marked asymmetry of the sperm whale head becomes apparent already at these early

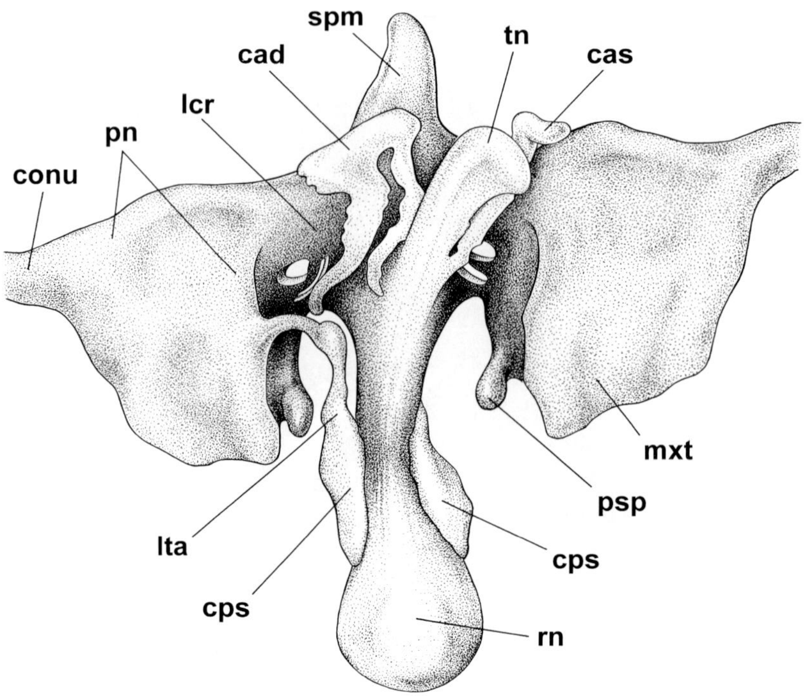

Fig. 18. *Physeter catodon* embryo, 150 mm TL. Frontal view of the cartilaginous structures of the embryonic nasal skull

developmental stages. The medial part of the paries nasi shows a typical inside furling of its outer edge, corresponding to the maxilloturbinale. The posterior or inferior part of the paries nasi ends in a slightly rounded cupula nasi posterior, the medial edge of which emits a short processus paraseptalis pointing anteriorly. A commissura orbitonasalis is present, however not as a single cartilaginous band as usual, but split into an upper and a lower branch. Even this is among the singularities of the sperm whale head. As far as the asymmetry is concerned, the posterior side wall structures are all somewhat more massively developed and reach farther ventrally on the right side than on the left.

The asymmetry of the anterior side wall structures is still more marked. On the left side, these structures consist of a deeply situated narrow bar which is not connected with the posterior side wall structures and whose anterior tip reaches fairly far towards the anterior region as a structure accompanying the rostrum nasi. On the right side, the anterior side wall structures consist of a highly situated and broad plate which is connected with the posterior side wall structures and produces anteriorly only a short and blunt tip that reaches only as far as the base of the rostrum nasi. On either side, the boundaries between the individual elements of the anterior side wall structures cannot be discerned unequivocally. Only a narrowing about the middle of the structures permits an approximation of the place in which the posteriorly situated lamina transversalis passes into the anteriorly situated cartilago paraseptalis. Any boundaries of the processus lateralis ventralis and of the cartilago ductus nasopalatini are quite invisible.

In general, the dermal bones in *Physeter* show the same connections with certain cartilaginous structures of the nasal skeleton as in all other toothed whales examined so far. From the upper lateral region, the praemaxillare accompanies the rostrum nasi and the anterior side wall structures up to behind the caudal tip of the lamina transversalis anterior. From below, the vomer accompanies the rostrum nasi and the anterior part of the anterior side wall structures, i.e., the cartilago paraseptalis, extending far posteriorly, down below the cupula nasi posterior of the posterior side wall structures. The maxillare, the largest of the three dermal bones, lies mostly lateral of the praemaxillare and the vomer and extends from the rostrum nasi down behind the posterior side wall structures.

However, the development of the dermal bones in *Physeter* differs conspicuously in several particulars from the situation found in the remaining toothed whales. First, it is the long lasting considerable arrest in their growth, compared to the much higher growth rate of the cartilaginous structures. This is particularly distinct in the parts situated rostrally. The arrest should be understood as a relative delay as compared to the development of the cartilaginous rostrum, as it is the task of the latter to initiate the growth processes of the extremely long anterior part of the head of the sperm whale. Second, it is the heavily altered form of the dermal bones. Their posterior sections become upturned and extended in the direction towards the tectum nasi. In this way they strengthen the basal part of the tectum nasi from which protrudes a narrow cartilaginous bar of the nasal roof cartilage, extending diagonally through the whole of the soft anterior part of the head. However, the cartilaginous bar itself is not accompanied by the dermal bones.

In addition, the dermal bones of the sperm whale show distinct asymmetry, the same as the cartilaginous structures which they accompany. Particularly their posterior sections are situated higher up on the right side and they extend more caudally than the corresponding structures on the left side of the head.

4.2
Mysticeti

4.2.1
Fin Whale *Balaenoptera physalus*

An embryo 156 mm in size was available; it was comparable with the embryonic *Stenella* in Group V. The nasal skull of this representative of the baleen whales, Mysticeti, while agreeing in its basic structure with the situation found in *Stenella* and other toothed whales, Odontoceti, differs from it in several particulars (Figs. 19-22). On the whole, the nasal skull is lower and longer. The rostrum is conspicuously slender. The anterior side wall structures do not extend so much anteriorly. The posterior side wall structures are not so posteriorly shifted, they are not so heavily flattened and their position is not perpendicular but forming an angle of about 45° with the longitudinal axis of the skull. On its inner surface, several elements of the primary nasal capsule of terrestrial mammals can be discerned.

As in all species examined so far, the medial structures form a longitudinal vertical plate. They form a continuum in which, in front, the rostrum nasi, and behind, the septum nasi, and still more posteriorly, the septum interorbitale, a transition to the neurocranium, can be distinguished. The plate is triangular, its tip pointing anteriorly and its highest place posteriorly. While the ratio of the length measured from the rostral tip to the end of the septum nasi, to the maximum height in the region of the tectum nasi is about 2:1 in the comparable stages of *Stenella* as well as in most other Odontoceti, in *Balaenoptera* this ratio is about 4:1, that is, twice as high.

The posterior edge of the median structures falls off only gently towards the posterior region. In *Stenella* and many other Odontoceti, this edge is almost perpendicular to the longitudinal axis of the nasal skull, i.e., at an angle of about 90°, whereas in *Balaenoptera* the angle is about 45° only. The narrow spina mesethmoidalis points steeply upwards and attains a considerable length of about 50% of the maximum height of the septum nasi. Immediately in front of the spina there is a deepened furrow, the sulcus supraseptalis.

The posterior side wall structures are fairly massive. They arise from the upper edge of the septum nasi, pointing to the left and right, and forming two powerful furls opened ventrally.

Viewed from the outside (Figs. 19, 20), the following elements can be distinguished. In front, the cupula nasi anterior, protruding far upward and richly articulated. In the middle top, the tectum nasi, passing in the posterior ventral direction into the paries nasi without any visible boundary. The tectum nasi emits, anteriorly, a small process, the processus paranasalis. At the lower edge of the paries nasi, there is another process, the processus maxillaris posterior. In the dorsal posterior direction, the paries nasi passes into a powerful commissura orbitonasalis, receiving a connection with the ala orbitalis. This connection is considerably more massive than in the Odontoceti. The cupula nasi posterior closes the furl of the posterior side wall structures in a posterior ventral direction. The whole furl lies at the same level with the caudal edge of the septum nasi, i.e., it falls off at an angle of about 45° posteriorly and ventrally.

Viewed from the inside (Figs. 21, 22), a complicated relief of the nasal capsule proper is visible. First of all, let us follow the main nasal cavity in the direction from anterior towards posterior. The entrance is formed by the rim of the paries nasi, situated below the cupula nasi anterior and the tectum nasi. This entrance ends in a semicircular band open towards the posterior side, the crista semicircularis. Lateral from it lies the most anterior accessory cavity, the recessus frontoturbinalis. Behind it the frontoturbinale projects as a moderately arched band. Behind it there is another accessory cavity, the recessus maxillaris. Thereupon follows a powerful process protruding far to the inside and connected with the paries nasi through just a narrow shaft. With considerable certainty, this process can be interpreted as the ethmoturbinale I. Behind it lies the largest accessory cavity, the recessus ethmoidalis. The wide vault of the cupula nasi posterior with a short recurrent processus paraseptalis posterior narrows this last accessory cavity, at the same time forming the posterior rim of the whole nasal capsule. The lower edge of the nasal capsule is furled like the maxilloturbinale, reaching from the crista semicircularis in front down to the processus paraseptalis posterior in the rear. The upper part of the nasal capsule is formed by the lamina cribrosa opening towards the neurocranium through a large foramen olfactorium.

The anterior side wall structures are relatively small. Only two elements can be distinguished among them with certainty. In the rear lies the narrow lamina

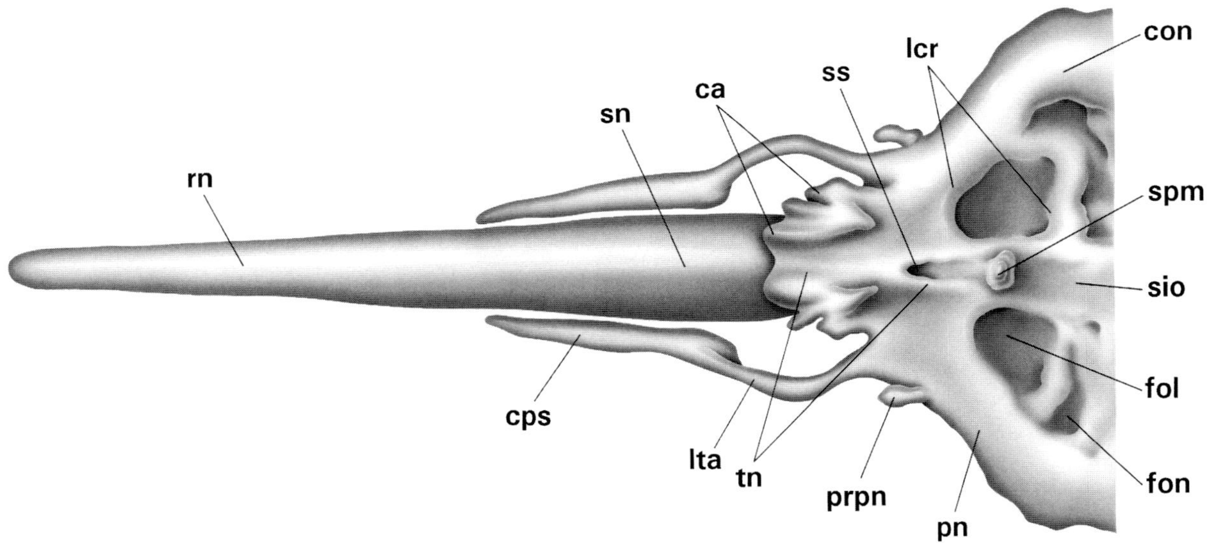

Fig. 19. *Balaenoptera physalus* embryo, 156 mm TL. Dorsal view of the cartilaginous structures of the embryonic nasal skull

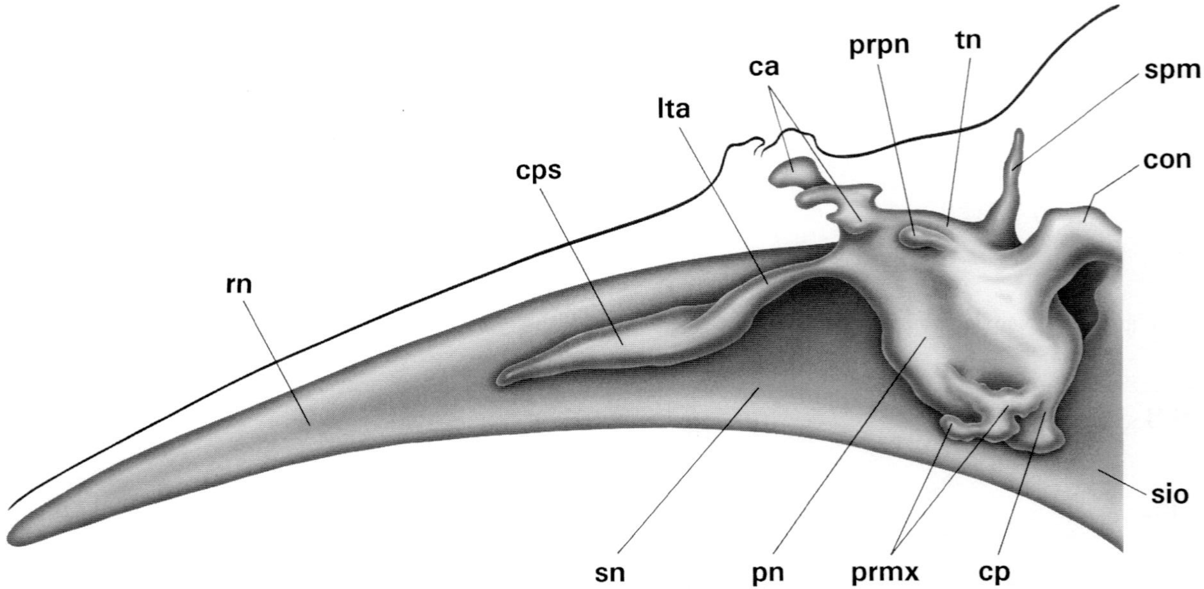

Fig. 20. *Balaenoptera physalus* embryo, 156 mm TL. Lateral view of the cartilaginous structures of the embryonic nasal skull

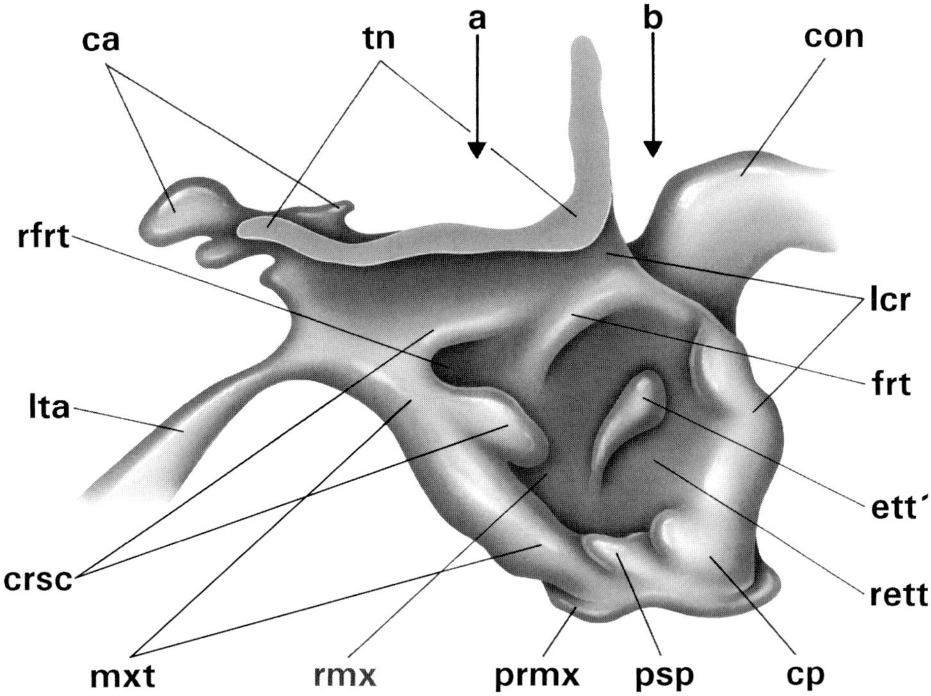

Fig. 21. *Balaenoptera physalus* embryo, 156 mm TL. The right nasal cavity viewed from the inside. The *arrows* indicate the section planes in Fig. 22

transversalis anterior, representing the connection with the posterior side wall structures. It originates from the edge of the paries nasi and runs in a wide lateral arch anteriorly and ventrally. In front lies the cartilago paraseptalis which is somewhat more massive than the lamina transversals anterior. It is intimate with the rostrum nasi like a laterally flattened plate. Its anterior tip shows a slight protrusion, probably the integrated material of the cartilago ductus nasopalatini. There is no processus paraseptalis anterior.

The dermal bones show the same topographic relationships as in Odontoceti. The praemaxillare accompanies, from a lateral direction, the upper edge of the rostrum nasi and the anterior side wall structures all along their length up to the paries nasi. The maxillare accompanies, from laterally, the praemaxillare, reaching beyond the posterior side wall structures. The vomer forms a through cover of the lower edge of the median structures and extends from the rostral tip up to the septum interorbitale. Its upper edge supports the cartilago paraseptalis as well as the processus paraseptalis posterior.

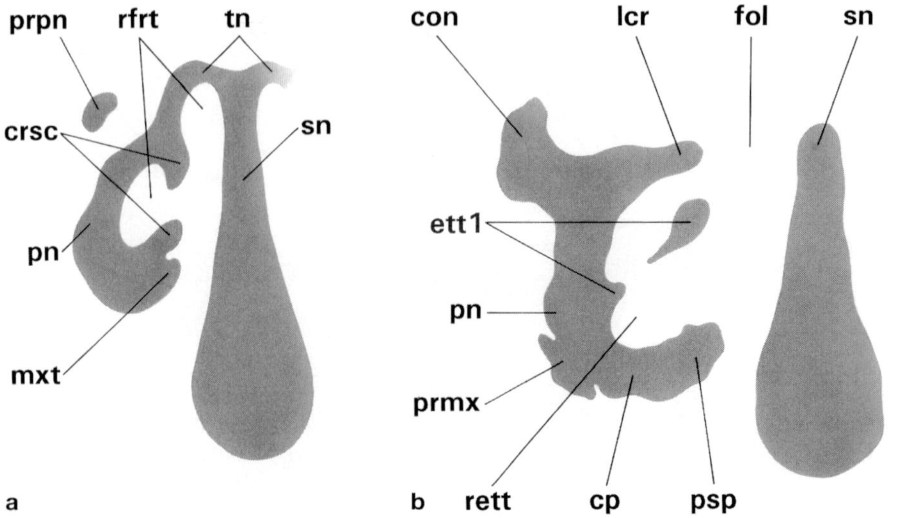

Fig. 22a,b. *Balaenoptera physalus* embryo, 156 mm TL. The right nasal cavity shown in cross sections. **a** and **b** correspond to the section planes indicated by arrows in Fig. 21

4.2.2 Minke Whale *Balaenoptera acutorostrata*

An older embryo of this species, 445 mm in size, was available. It was examined only macroscopically. A report on the same species had already been presented by Kükenthal (1893) who examined an embryo 201 mm in size, and later by De Burlet (1914a) whose findings in an embryo 105 mm in size were also adopted by De Beer (1937). The macroscopic examination provided only a rough outline of the form and proportions of the structures of the cartilaginous nasal skull, as well as information on the mutual relationships between the chondrocranium and the dermatocranium.

In general, the dermal bones have increased in relative size at this advanced developmental stage, whereas the cartilaginous skull parts have decreased in relative size. The external shape of the nasal skull is solely determined by the dermal bones. The cartilaginous structures, except for the rostrum nasi, have been compressed to the inside, as they are only relatively small parts of the skeleton. Their mutual connections have been preserved, as observed in the younger stage of *Balaenoptera physalus*. The praemaxillare is connected with the cartilaginous rostrum nasi and with the anterior side wall structures. The maxillare, lying laterally therefrom, is connected with the posterior side wall structures in its posterior region. The vomer is connected with the cartilaginous rostrum nasi as well as with the processus paraseptalis posterior.

Among the cartilaginous elements, the rostrum nasi shows the most advanced development and is the only cartilaginous structure that keeps step with the high growth rate of the dermal bones. It forms a powerful partition between the dermal parts of the maxilla, reaching together with it up to the extreme tip of the maxilla. The posterior end of the rostrum nasi continues above in the tectum nasi and ends in a narrow spina mesethmoidalis propped upwards. The posterior side wall structures, observed from the outside, show much the same shape and inclination as those of the younger stage of *Balaenoptera physalus*. The relief of their inner surface has not been examined. From the posterior side wall structures, from the anterior edge of the paries nasi, the anterior side wall structures arise anteriorly in the form of a narrow arched bar. In their posterior part, the lamina transversalis anterior, and in their anterior part, the cartilago paraseptalis are distinguishable.

4.2.3 Humpback Whale *Megaptera novaeangliae*

Serial sections of three embryos were available, 45, 90 and 107 mm in size. The two larger embryos are approximately comparable with the embryonic *Stenella* in Group IV. The development of these stages has advanced to the extent that it is also directly comparable with that of the somewhat bigger embryo of *Balaenoptera physalus* of 156 mm. The skull development of *Megaptera novaeangliae* had been examined by Honigmann (1915, 1917) in an embryo 95 mm in size. His observations were later adopted by De Beer (1937).

But for relatively small differences, the nasal skull of *Megaptera* resembles that of *Balaenoptera*. It has largely the same shape, about the same proportional relationships, as well as the same mutual inclination or angles of the individual structures (Figs. 23–26).

The median structures show the same division as in *Balaenoptera*. The only difference worth mentioning is that the rostrum nasi appears to be somewhat shorter and higher, which is apparently due to the fact that the stage of *Megaptera* is a little younger.

The posterior side wall structures (Figs. 23, 24) show, in lateral view, a somewhat differently shaped cupula nasi anterior and a slightly simpler division of the side wall than in *Balaenoptera*.

There is no processus maxillaris posterior. In the inner relief the same structures are discernible yet somewhat different in shape (Figs. 25, 26). In the anterior-posterior direction they include: the anterior arch of the paries nasi, the crista semicircularis, the recessus frontoturbinalis, the frontoturbinale, the recessus maxillaris, the ethmoturbinale I, the recessus ethmoturbinalis and, finally, the cupula nasi posterior with the processus paraseptalis posterior. The lower edge of the nasal capsule consists of the furl of the maxilloturbinale. The upper edge consists of the tectum nasi and the lamina cribrosa with a large foramen olfactorium.

The anterior side wall structures reach, with their anterior tip, almost down to the lower edge of the rostrum nasi. That is the most conspicuous difference. Otherwise they are exactly the same shape as in *Balaenoptera*, and they also consist of only two distinct elements, the lamina transversalis anterior and the cartilago paraseptalis.

The dermal bones show a conspicuous linkage with the same cartilaginous structures of the nasal skull as in all Odontoceti examined, as well as in the Mysticeti of the genus *Balaenoptera* examined. The praemaxillare accompanies the rostrum nasi and the anterior side wall structures. The maxillare is situated laterally therefrom and reaches beyond the posterior side wall structures. The vomer accompanies the rostrum nasi and the septum nasi and supports from below the cartilago paraseptalis as well as the processus paraseptalis posterior.

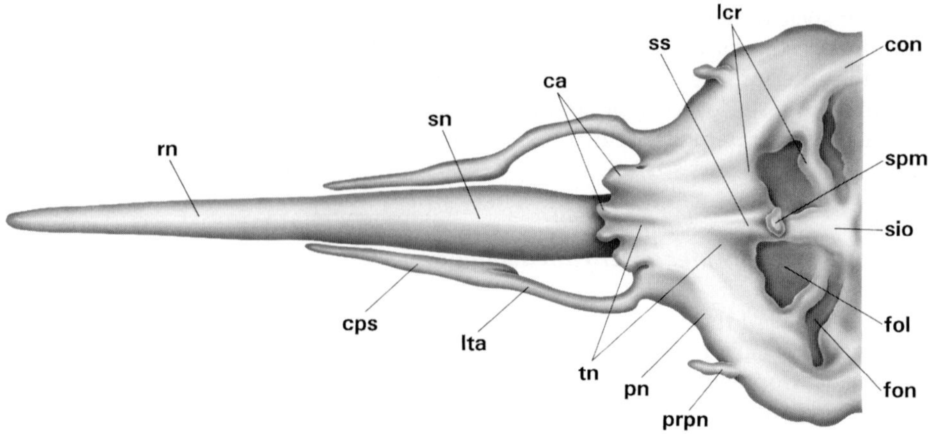

Fig. 23. *Megaptera novaeangliae* embryo, 107 mm TL. Dorsal view of the cartilaginous structures of the embryonic nasal skull

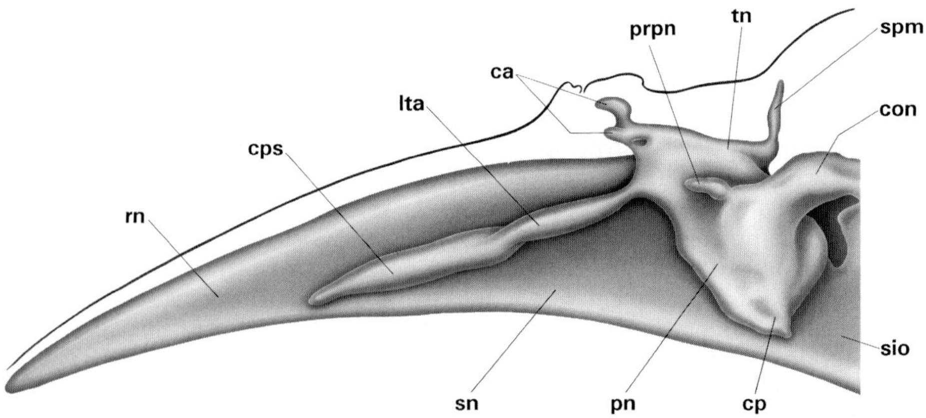

Fig. 24. *Megaptera novaeangliae* embryo, 107 mm TL. Lateral view of the cartilaginous structures of the embryonic nasal skull

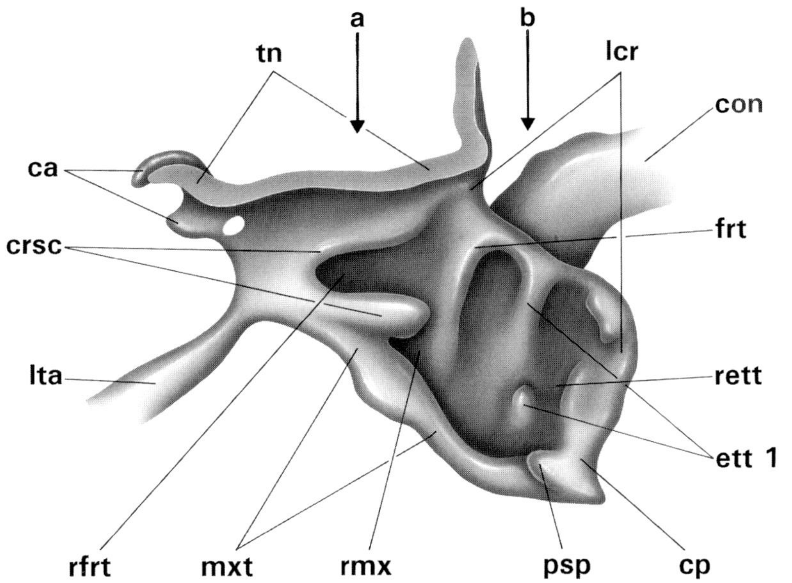

Fig. 25. *Megaptera novaeangliae* embryo, 107 mm TL. The right nasal cavity viewed from the inside. The *arrows* indicate the section planes in Fig. 26

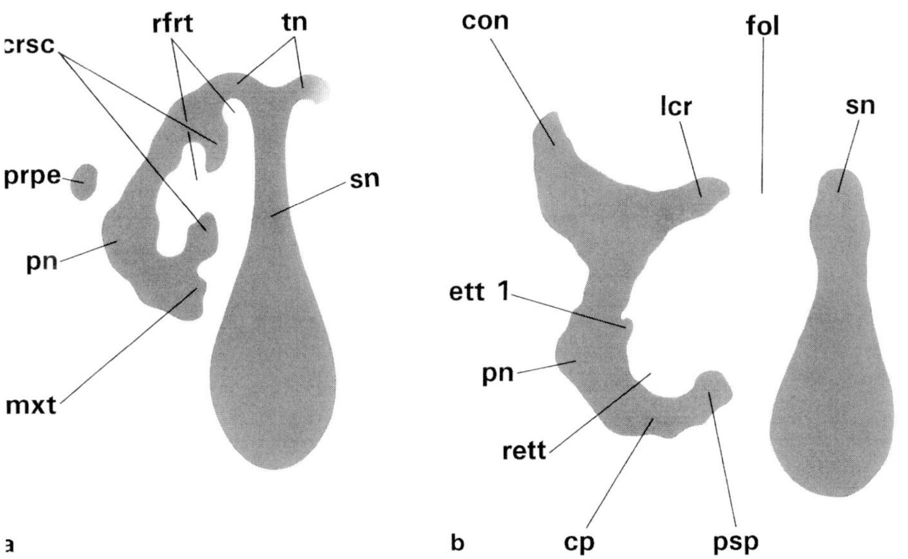

Fig. 26a,b. *Megaptera novaeangliae* embryo, 107 mm TL. The right nasal cavity shown in cross sections. **a** and **b** correspond to the section planes indicated by the arrows in Fig. 25

5 Metamorphosis and Translocation of the Nasal Structures

The almost continuous developmental series of embryonic *Stenella*, *Delphinus*, *Phocoena* and *Physeter* served to follow the course of early morphogeny of the nasal skull as a continuous process (Figs. 27–30). Only in this way are the transformations and displacements of the individual structures of the nasal skeleton of cetaceans understandable, and their homologization with the corresponding structures in terrestrial mammals can be demonstrated.

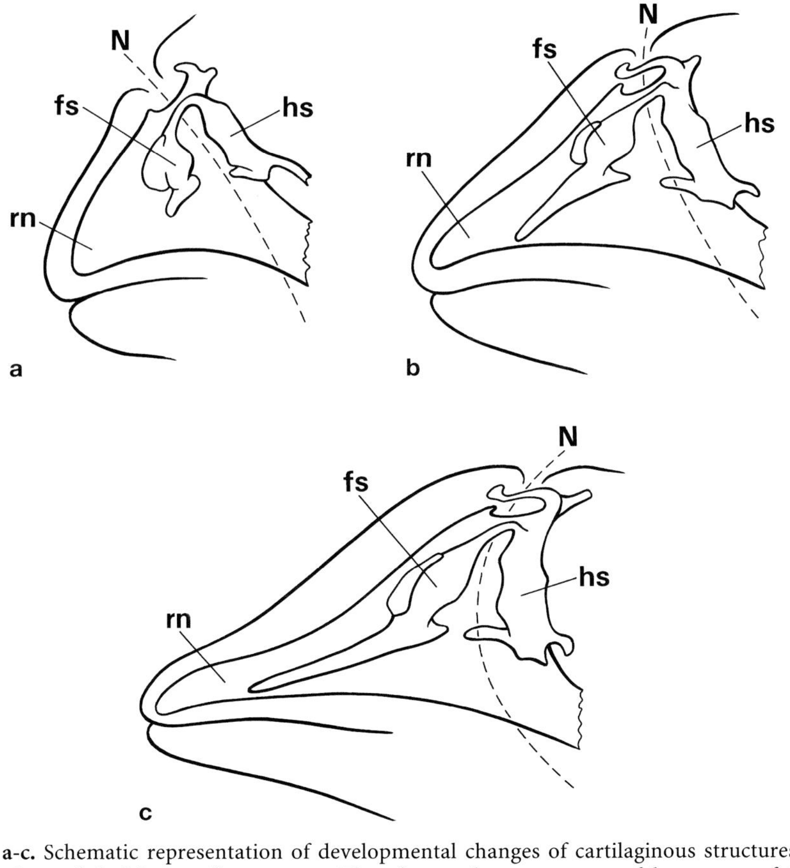

Fig. 27 a-c. Schematic representation of developmental changes of cartilaginous structures of the embryonic nasal skull in *Stenella attenuata*. The schemes show a sequence of three stages of about 40 (**a**), 70 (**b**) and 150 mm TL (**c**)

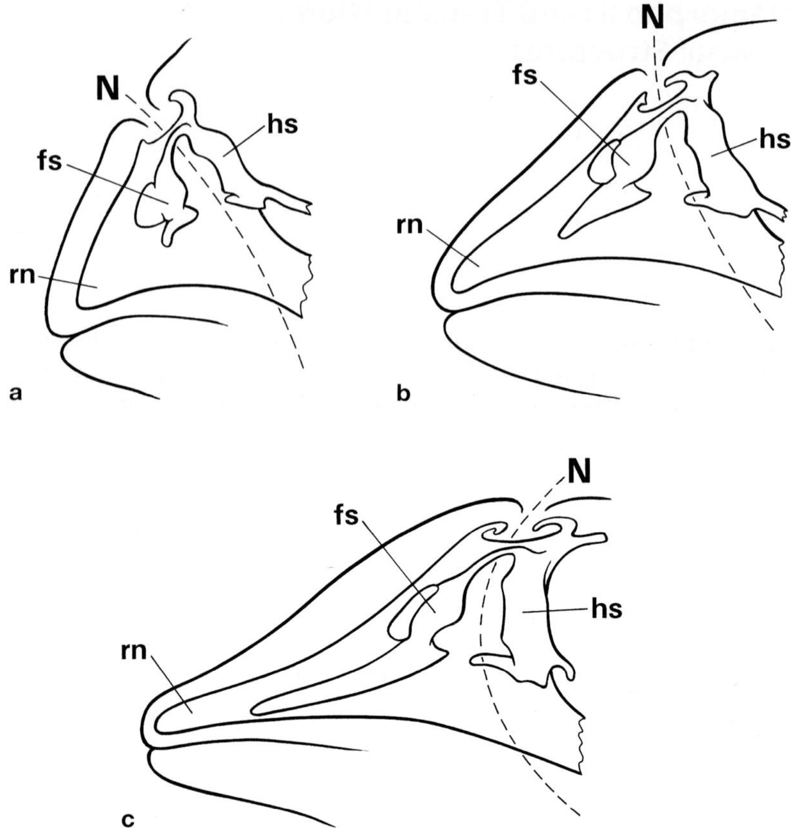

Fig. 28a-c. Schematic representation of developmental changes of the cartilaginous structures of the embryonic nasal skull in *Delphinus delphis*. The schemes show a sequence of three stages of about 40 (**a**), 80 (**b**) and 130 mm TL (**c**)

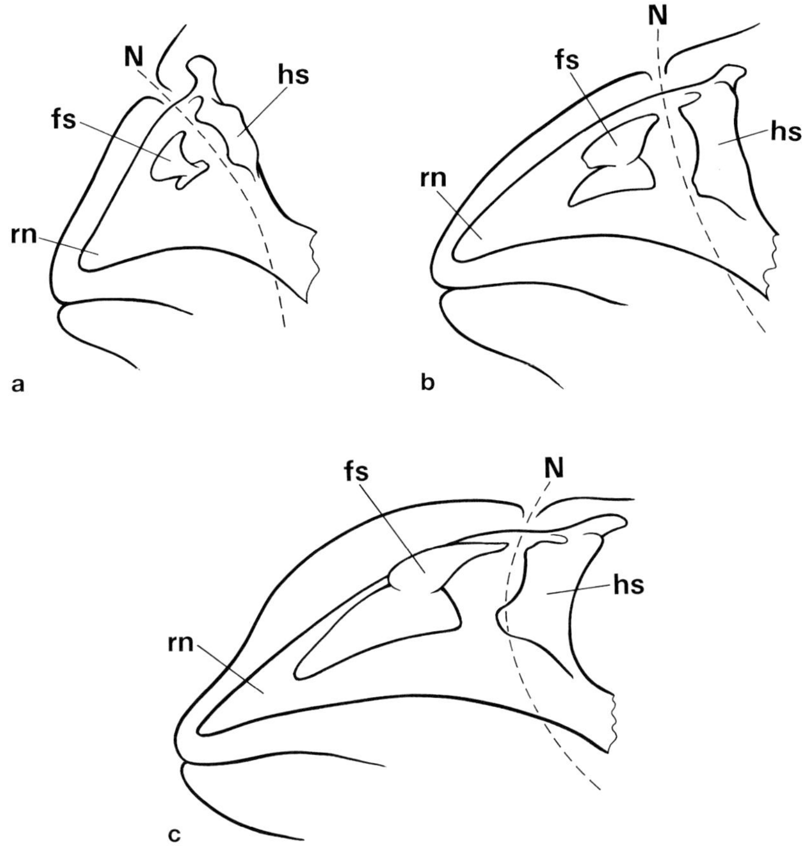

Fig. 29a-c. Schematic representation of developmental changes of the cartilaginous structures of the embryonic nasal skull in *Phocoena phocoena*. The schemes show a sequence of three stages of about 40 (**a**), 90 (**b**) and 170 mm TL (**c**)

5.2 Stages Around 80 mm in Size

At these medium-sized developmental stages (Figs. 27–30, part b), a similarity with the nasal skull of terrestrial mammals can hardly be detected. The median wall dividing the nasal cavity becomes distinctly triangular in shape. Its tip, slightly bent downward, projects in a long rostrum nasi. The posterior part of the triangle, consisting of the septum nasi proper, increases considerably in height. Its upper tip ends free in the slightly recurred spina mesethmoidalis and its inferior tip continues into the broadened septum interorbitale. The initially horizontal or slightly oblique upper edge of the septum nasi becomes erect and falls off steeply towards the posterior edge. In this way it forms the posterior closure of the septum nasi.

At the same time, the nasal opening moves away from the tip of the snout and rests on the forehead. The two nasal passages assume an increasingly vertical position, forming an angle of about 70°–80° with the longitudinal axis of the skull. They are parallel to the posterior edge of the septum nasi and are accompanied by the posterior side wall structures, i.e., by the initial complex of the nasal roof and the lateral wall structures. The elements of this complex, as already observed at the earliest stages, are additionally joined by the distinctly developed maxilloturbinale in the medial section and an indistinctly defined elevation in the inferior one, which can probably be considered to be the ethmoturbinale I. The processus paraseptalis posterior increases in size and projects forwards.

The anterior side wall structures change their shape and position even more than the posterior ones. While the narrow bridge is preserved between the two, the anterior side wall structures separate from the initial nasal floor and turn anteriorly, parallel to the rostrum nasi. The processus lateralis ventralis remains relatively small, whereas the lamina transversalis anterior and, above all, the cartilago paraseptalis increase in size quite considerably. Their shape is fundamentally altered. As a whole, they form a very long and narrow band which no longer serves as a nasal floor but becomes, to a certain extent, a component of the rostrum.

The homology of the displaced and transformed cartilaginous structures with those of the initial nasal skull of terrestrial mammals is also confirmed by their connection with certain desmal elements. At these medium-sized stages the anlagen of the desmal skull bones are already well developed and they join the cartilaginous structures laterally. At the same time, the firm connections between the praemaxillare and the lamina transversalis anterior, between the vomer and the cartilago paraseptalis and the processus paraseptalis posterior are preserved, the same as is the case with the initial nasal skull of terrestrial mammals.

The relationships described above are true of *Stenella, Delphinus* and, except for several negligible particulars, also *Phocoena.* While the transformation of the nasal skull takes the same course even in *Physeter*, in addition to that, already at this developmental stage, a process rises from the tectum and septum nasi, the later "nasal roof cartilage" of the adults (For particulars, see Sect. 4.1.7 on *Physeter*).

5.3
Stages Around 150 mm in Size

At these older developmental stages (Figs. 27–30, part c), the morphogeny of the nasal skull is largely completed. The transformation and translocation displacement of the individual elements has basically ended. All embryonic nasal structures of the chondrocranium as well as those of the dermatocranium are fully developed and have assumed their final position and shape typical of the nasal skull of cetaceans.

The septum nasi lengthens far rostrally and anteriorly, and ends in a narrow tip of the rostrum nasi. The posterior edge of the septum nasi is quite vertical and its upper end turns down with the spina mesethmoidalis. At the same time, the nasal opening moves still farther from the snout tip and rests at a posterior superior position, about the vertex of the head. The two nasal passages run quite perpendicular to the longitudinal axis of the skull at an angle of about 90°, or they are even inclined more caudally. This is the same direction in which even the posterior edge of the septum nasi is posited, the same as the posterior side wall structures accompanying the nasal passages. The anterior side wall structures become still more separated from the nasal passages and functionally become a component of the rostrum nasi which they accompany far rostrally. Of all structures of the cartilaginous nasal skull, the anterior side wall structures have the most increase in size.

The desmal elements, the praemaxillare, maxillare and vomer, have still retained the same connection with the corresponding cartilaginous structures as initially at the medium-sized stages. They form a complete, massive case around the initial cartilaginous nasal structures. From now on, it is the desmal elements and the dermatocranium formed by them that determine the subsequent development of the nasal and maxillary skull. In it, no more transformations or changes in shape worth mentioning take place. The development to follow is chiefly characterized by intense growth.

The relationships described above are true of *Stenella, Delphinus* and *Phocoena.* Basically, this developmental mode is also true of *Physeter,* except for particulars pertaining to the development of the nasal roof cartilage of the sperm whale in connection with the development of the spermaceti organ, as described in detail in Chap. 7.4.

5.4
Course of the Most Important Transformations

Beginning with the youngest stage, the nasal skeleton of which presents a kind of transition from terrestrial mammals to whales and dolphins and in which the position of the structures largely reflects the initial relationships, the most significant subsequent transformation processes can be characterized as follows (Figs. 31–37).

The snout tip and the nostril become separated. The septum nasi runs out rostrally into a rostrum nasi reaching far anteriorly. In the posterior part of the septum nasi, the spina mesethmoidalis and the upper part of the septum nasi become shifted downwards so that finally the posterior edge of the septum falls off vertically.

The initial complex of the nasal roof and the side lateral wall, here summarized as the posterior side wall structures, become shortened and considerably reduced. Above

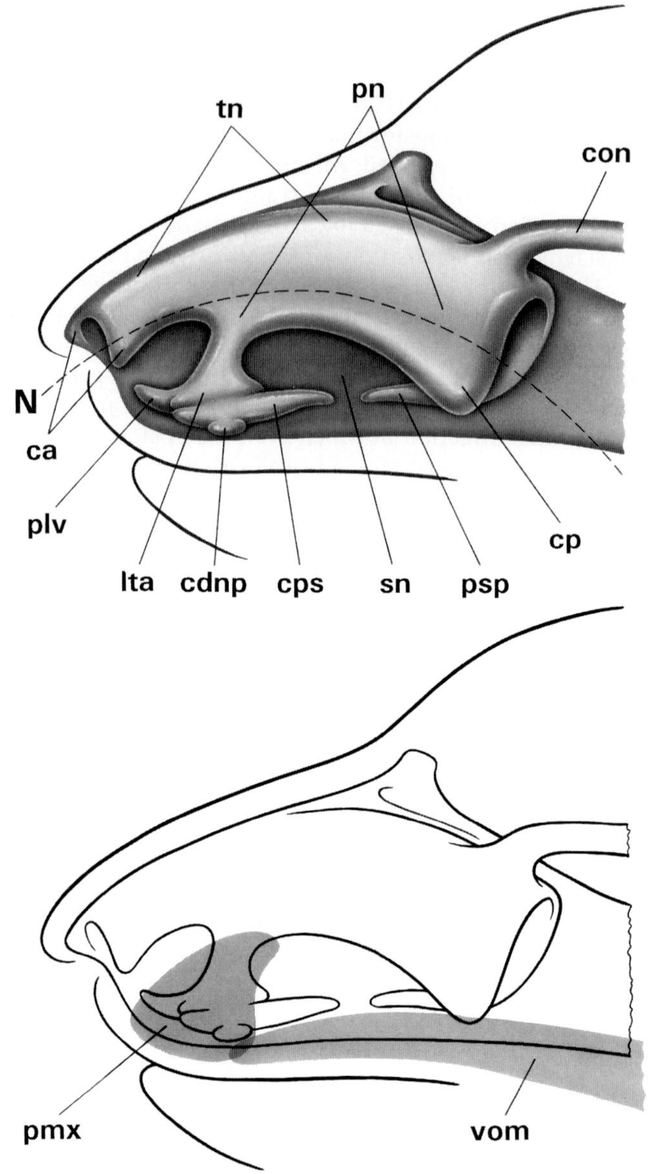

Fig. 31. Cartilaginous structures (*top*) and dermal bones (*bottom*) of a generalized embryonic nasal skull in terrestrial mammals

all, the reduction pertains to the tectum nasi, less so to the paries nasi. At the same time, the posterior side wall structures turn up about 90°. Their anterior part, consisting predominantly of the anterior nasal capsule, the cupula nasi anterior, is shifted in an occipital superior direction. Their posterior part, consisting predominantly of the posterior nasal capsule, the cupula nasi posterior, is shifted

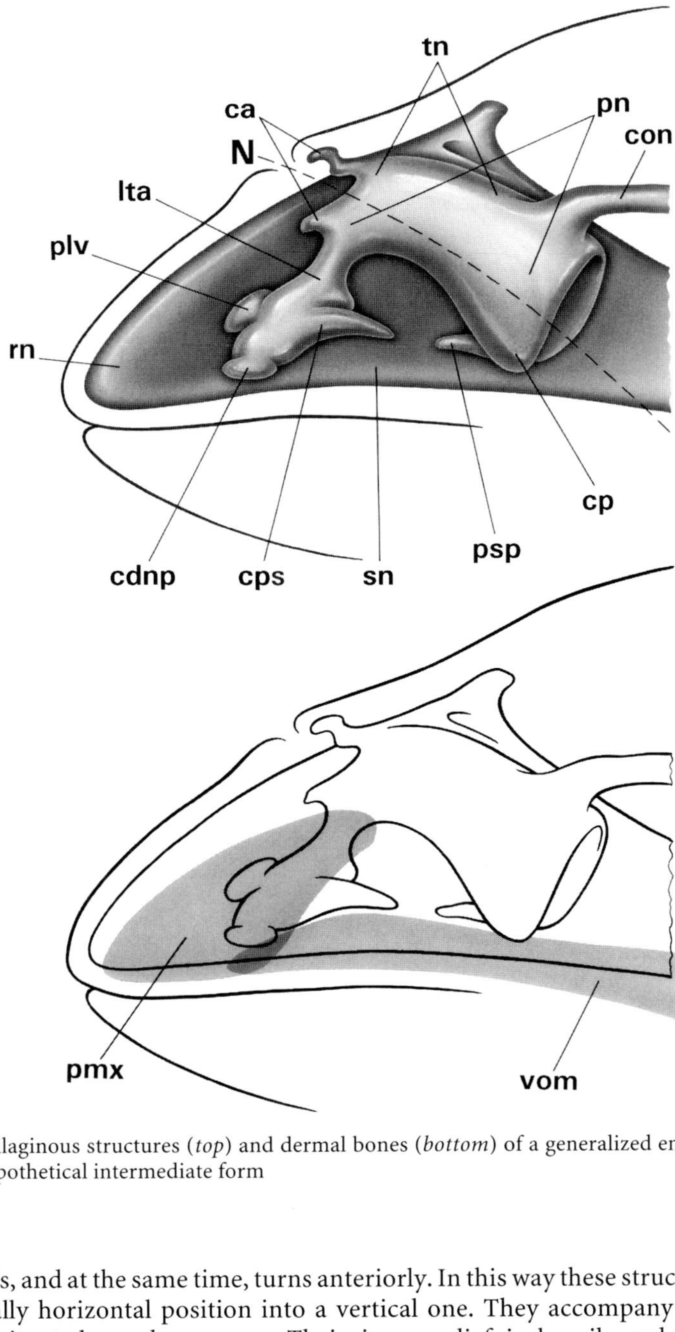

Fig. 32. Cartilaginous structures (*top*) and dermal bones (*bottom*) of a generalized embryonic nasal skull in a hypothetical intermediate form

downwards, and at the same time, turns anteriorly. In this way these structures change their initially horizontal position into a vertical one. They accompany the likewise vertically situated nasal passages. Their inner relief is heavily reduced. Of the conch-like structures, typical of the macrosmatic terrestrial mammals, only the maxilloturbinale and in part, the ethmoturbinale I are discernible. On the whole, the posterior side wall structures are heavily simplified and flattened. Together with the

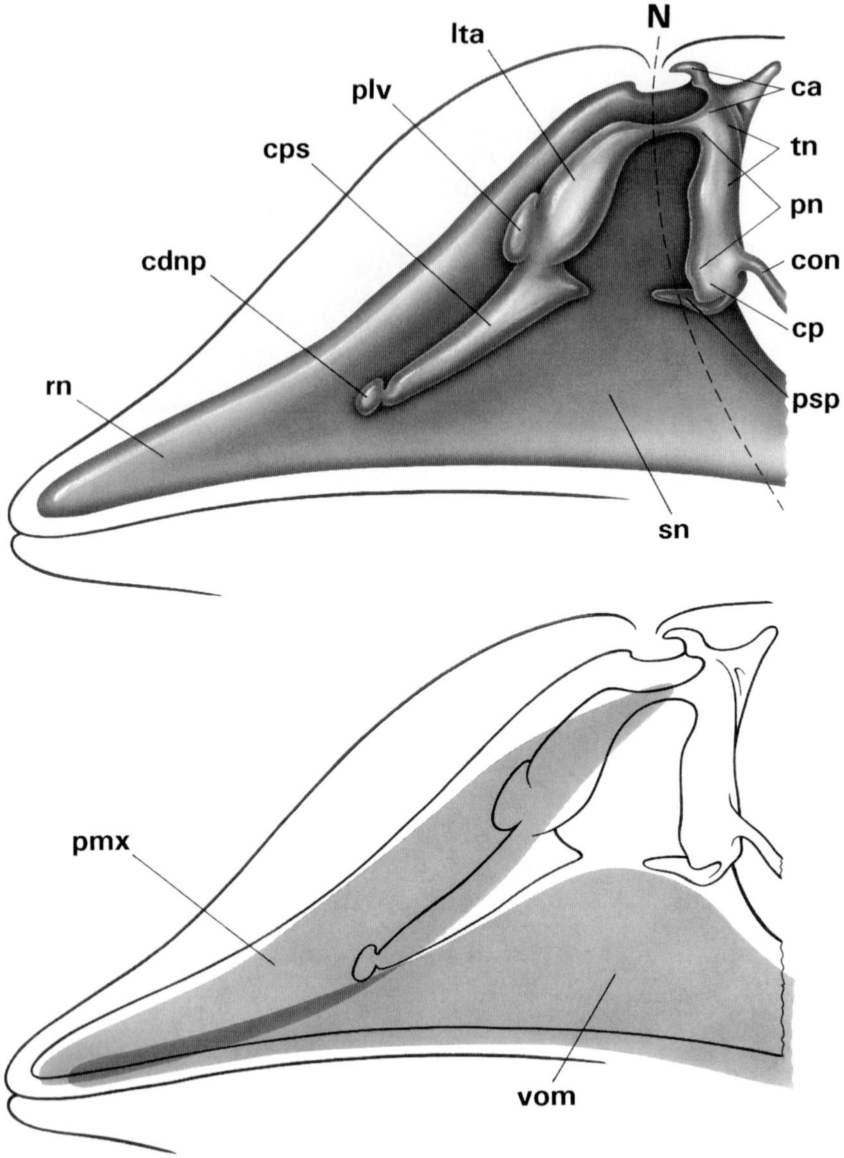

Fig. 33. Cartilaginous structures (*top*) and dermal bones (*bottom*) of a generalized embryonic nasal skull in cetacean of typus *Stenella*

vertical posterior edge of the septum nasi, they participate in the formation of the anterior wall of the neurocranium.

The elements of the initial nasal floor, here summarized as the anterior side wall structures, increase unusually in size and develop into an important part of the upper

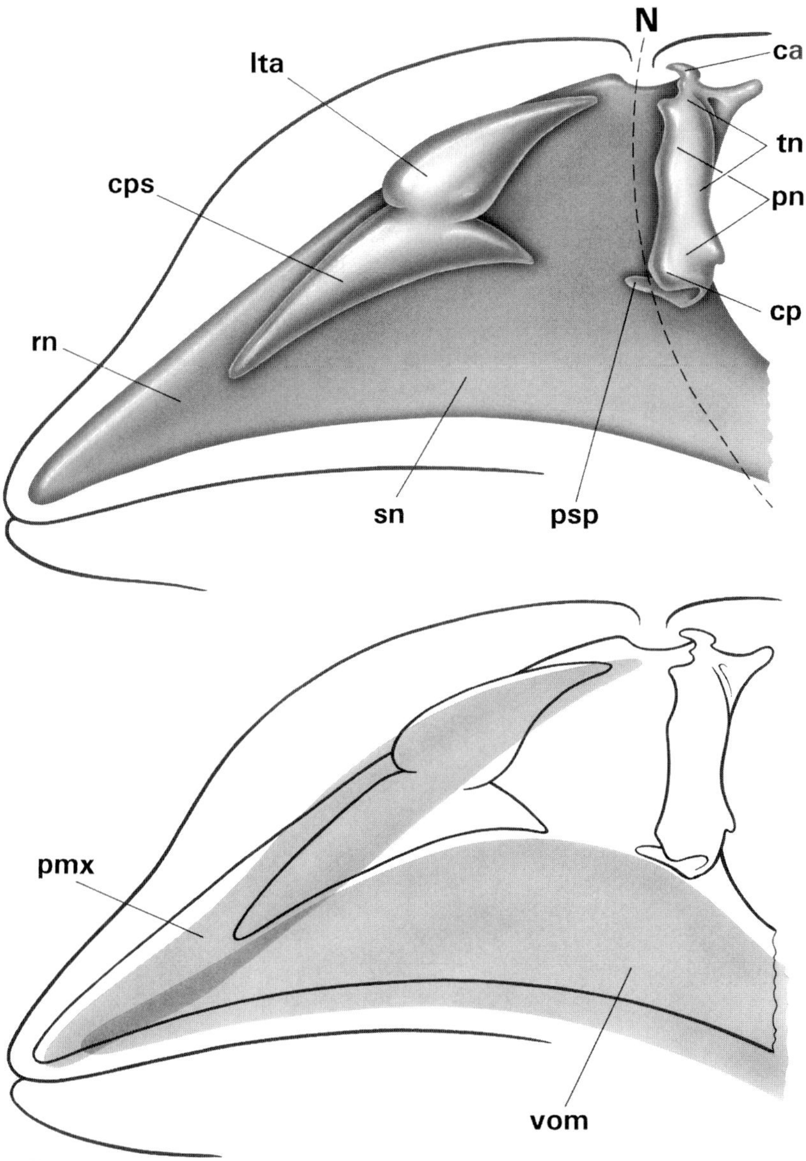

Fig. 34. Cartilaginous structures (*top*) and dermal bones (*bottom*) of a generalized embryonic nasal skull in cetacean of typus *Phocoena*

jaw. They separate from the nasal passages and shift anteriorly. At the same time, they turn by almost 180°, so that the cartilago paraseptalis, initially pointing towards the posterior, now protrudes in the opposite direction anteriorly, and at the same time, shifts in front the cartilago ductus nasopalatini. In the same manner, also the lamina transversalis anterior, and with it the processus lateralis ventralis are turned. The

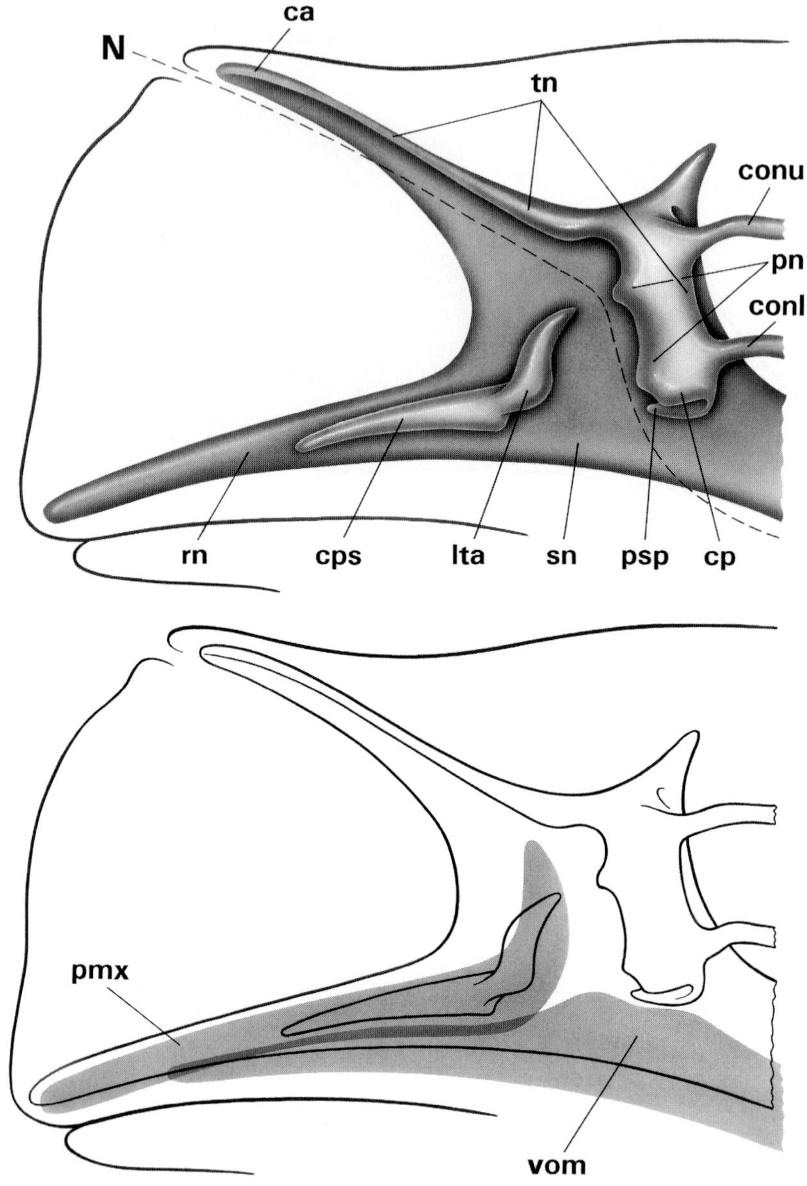

Fig. 35. Cartilaginous structures (*top*) and dermal bones (*bottom*) of a generalized embryonic nasal skull in cetacean of typus *Physeter*

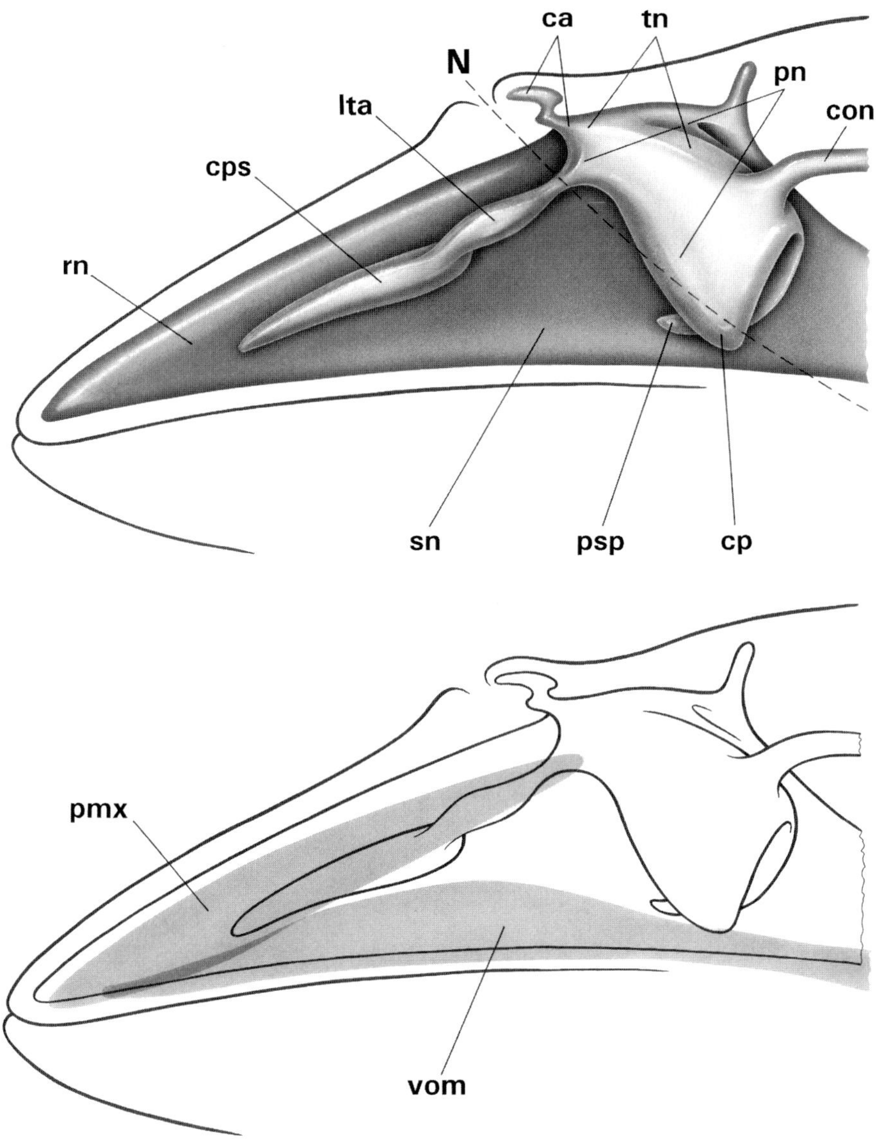

Fig. 36. Cartilaginous structures (*top*) and dermal bones (*bottom*) of a generalized embryonic nasal skull in cetacean of typus *Balaenoptera*

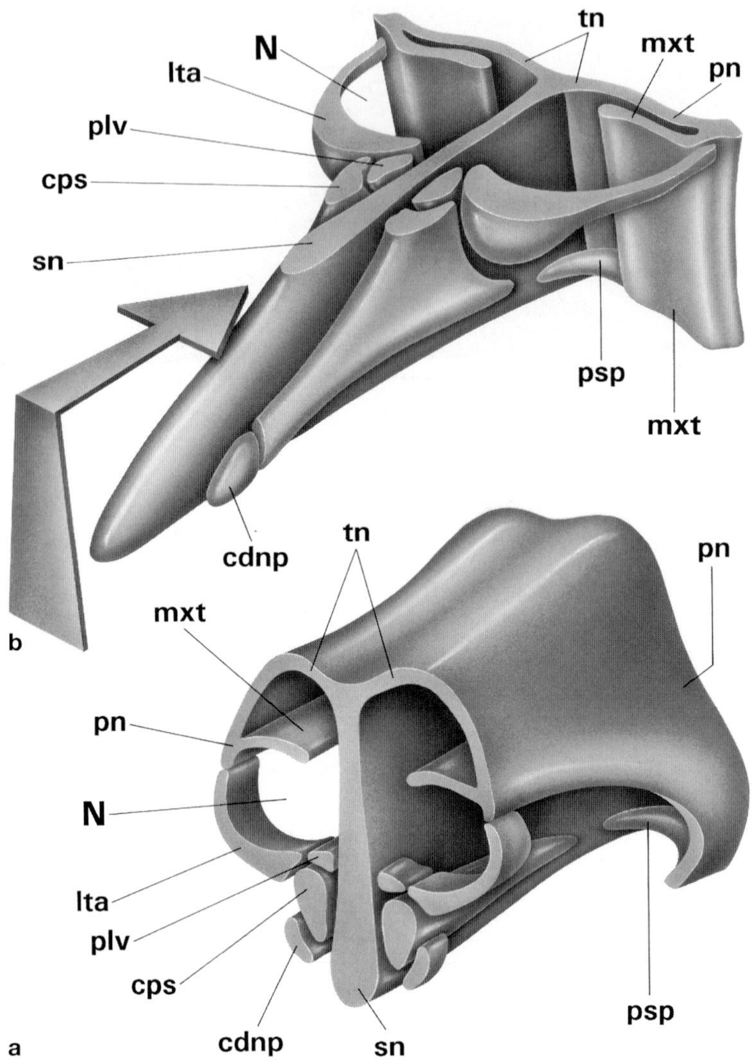

Fig. 37a,b. Comparison of the embryonic cartilaginous nasal structures in terrestrial mammals (**a**) and cetaceans (**b**). After the rotation of the nasal capsule through approximately 90°, both the transverse section in the terrestrial mammals and the horizontal section in cetaceans show the same generalized pattern

individual elements of the anterior side wall structures may occasionally become quite fused together, so that they reach anteriorly in the form of a single cartilaginous bar parallel to the rostrum nasi and participate in the development of the long snout typical of whales and dolphins.

6 Comparison of the Embryonic Nasal Structures

The exact knowledge of the continuous course of morphogenesis, described in the preceding chapters, is of extraordinary importance for the true interpretation of the individual structures of the heavily transformed cetacean nasal skull. Examinations made at individual stages, however exact, for the most part do not permit the recognition of natural connections and often lead to pure speculation. Thus, it is no surprise that even the earliest known attempt at such interpretation was accompanied by a grave error. In the place of the lamina transversalis anterior and the cartilago paraseptalis, Mayer (1835) thought he had found a rudimentary maxilloturbinale. His interpretation was later adopted by Kükenthal (1893) and Boenninghaus (1903) without any appreciable change, and it has been used even in recent times (Cave 1987, 1988; Praderi and Cave 1994) in attempts at clarifying the origin of the questionable Meckelian ossicles (see Sect. 7.5). Similarly, several contradictory and untenable interpretations have been presented especially for the anterior, less so for the posterior side wall structures (Kükenthal 1893; Abel 1902; Gruhl 1911; De Burlet 1913a,b, 1914a,b, 1916; Honigmann 1915, 1917; Schreiber 1916; Eales 1950). At the same time, the basic tendency of the transformation processes had already been correctly established, as indicated, above all, by the papers of De Burlet (1916) and Honigmann (1917). An attempt at a new evaluation of the hitherto known findings and at a unification of the different interpretations was made by De Beer (1937) in his monograph of the development of the mammalian skull. Not until much later was it possible to study the transformation processes and translocations of the nasal structures in a continuous course, first in *Phocoena* (Klima and Van Bree 1985) and then also in *Physeter* (Klima 1987a). Those observations, together with the findings in the developmental series of *Stenella* and *Delphinus* described in the present paper, provide a basis for a (hopefully) safe homologization of the individual nasal structures.

An unequivocal interpretation of the structures of the embryonic nasal skull is not always easy, not even in land mammals. A number of problems still awaiting solution are encountered – as pointed out, e.g., by Sturm (1936), Reinbach (1952), Slabý (1960) and others – in those mammals, the nose of which largely corresponds with the general pattern of the mammalian nose. However, an objection could be raised, stating that there is no general pattern of a mammalian nose at all and that specific modifications occur in any systematic group, at least down to the generic level, reflecting both evolutional influences and functional adaptations. Such modifications, however, are never as weighty as in the wholly unusual cetacean skull, and they can be almost invariably derived from an original common scheme. Despite the publication of several monographs reviewing the morphogeny of the mammalian skull (Matthes 1921a, 1921b, 1923; Stadtmüller 1936; De Beer 1937; Romer 1956; Starck 1967, 1979;

Moore 1981), many questions have remained unanswered, most of which pertain to the nasal skeleton. At the same time, there is no lack, up to the present, of detailed papers treating the morphogeny of the mammalian skull from the point of view of comparative anatomy (Sturm 1936; De Beer and Fell 1936; Starck 1941, 1943, 1960; Frick 1954; Frick and Starck 1963; Kuhn 1971; Köhnke 1980, 1985; Zeller 1983; Schrenk 1989; etc.). Many authors have paid special attention to the nasal structures (Nordenbos 1905; Gaupp 1908; Voit 1909; Frets 1912, 1913, 1914; Terry 1917; Sturm 1936; Reinbach 1952; Frick and Heckmann 1955; Slabý 1960, 1986, 1987b; Schliemann 1966, 1987, 1991; Maier 1980, 1991; Zeller 1987, 1989; Zeller et al. 1993; etc.). However, it is beyond the scope of the present paper to study in detail the morphogeny of the nasal skull in land mammals. For this reason, only such connections will be discussed below that are of immediate significance for the comparison with Cetacea.

The division and sequence of the structures discussed below corresponds to the natural division of the cetacean embryonic skull (see Chap. 3). Since the dermal bones occur rather late in ontogeny and since they are understood to be mere accompanying elements of the cartilaginous structures, they are not discussed separately in this paper, but directly together with the respective cartilaginous structures.

6.1 The Median Structures

6.1.1 Septum nasi

The vertical nasal septum consists of an cartilaginous plate which divides the regio ethmoidalis along the median plane into two bilaterally identical halves. According to Slabý (1960), the septum nasi is the most conservative element of the nasal capsule. In the course of ontogeny, it is the first element of the nasal skeleton to appear. For the most part, it develops through the fusion of the medial walls of the originally divided paired nasal capsules and parts of the trabeculae (Matthes 1921b; De Beer 1937; Kuhn 1971). This paired origin can rarely be demonstrated as pre-cartilaginous anlagen in the early stages of mammals, e.g., in some Rodentia (*Erethizon*: Struthers 1927; *Cavia*: Ratjova 1972a). At later stages the original paired character may be suggested by a longitudinal fissure-like sulcus ventralis on the lower edge of the septum. Still more distinctly visible is the paired origin of the septum in the sulcus supraseptalis, which can be found along the upper edge of the septum between the wings of the tectum and which is present, in many land mammals, in the form of a deep gape. According to De Beer (1937), even the paired paraseptal cartilages can be considered vestiges of the paired medial walls. Likewise, the cartilaginous septoturbinalia, present in some land mammals (e.g., Primates, *Papio*: Reinhard 1958, *Tarsius*: Starck 1982, 1984, *Galago*: Warich 1986; Xenarthra, *Dasypus*: Reinbach 1953, *Zaedus*: Reinbach 1955; Rodentia, *Lemmus*: Schrenk 1989) suggest the paired origin of the septum. According to Struthers (1927) and Spatz (1963, 1964), the material of the basally situated trabeculae should fuse to form the incrassate ventral edge of the septum.

The basal part of the septum nasi continues caudally into the septum interorbitale, the latter again passing into the trabecular plate. In this way, the base of the septum nasi becomes a component of the median longitudinal axis ("central stem," Fawcett) bearing the skull basis. The morphogeny as well as the maturation processes proceed from occipital to rostral regions. The anterior part of the septum, the rostrum nasi, is the last to develop. The whole cartilaginous plate of the septum nasi, ringing upwards from the basal axis, is irregularly quadrilateral in most land mammals. The anterior, rostral edge of this quadrangle is rather low. Its posterior, caudal edge, situated below the spina mesethmoidalis, mostly reaches the highest point of the septum. The anterior edge remains free, the same as the lower part of the posterior edge. The upper part of the posterior edge passes bilaterally into the lamina cribrosa. The upper edge assumes a connection with the tectum nasi. The lower edge was originally connected with the elements of the solum nasi but this connection disappears largely in most mammals. In many land mammals there is, in the rostral part of the septal plate, a transverse opening, the fenestra internasalis, besides other perforations (Starck 1941; Kuhn 1971).

Of the covering dermal bones, a close connection with the septum nasi is found, first of all, in the vomer, which embraces the lower edge of the septum in a U-shaped manner and accompanies it all along its length. In places, also the praemaxillare and the maxillare may fairly closely approach the septum to various extents.

A paired origin of the septum nasi at the pre-cartilaginous stage has been observed in none of the cetacean species examined. This may be due to the fact that the septum plays a quite special part in the formation of the unusually long snout of dolphins and whales. While the septum ranks among the earliest elements of the nasal capsule even in the Cetacea, the robust development and conspicuous changes in the shape of this structure imparts to it the characteristics of a new structure in which the ancient plesiomorph characteristics are suppressed and the new neomorph ones predominate. Thus, the septum nasi manifests itself already at the earliest stages as an unpaired anlage which, as the continuation of the "central stem", shows a rather early progressive growth rostrally. There are no septoturbinalia, nor a sulcus ventralis. A sulcus supraseptalis is only found in a heavily shortened form in the Mysticeti, *Balaenoptera* and *Megaptera*.

In the Cetacea (Fig. 38) the septum nasi forms a massive dividing wall, the dimensions of which, in contrast to those of the other parts of the nasal capsule, are much greater than in all other mammals. It consists of a vertical, median situated cartilaginous plate, triangular in shape. Its anterior tip projects in a long rostrum, its posterior upper tip reaches slightly beyond the caudal edge in the form of a spina mesethmoidalis, its posterior lower tip forming a gradual transition into the septum interorbitale. In cross section, the lower edge is somewhat rounded and considerably thicker than the narrow upper edge.

Compared to the nasal capsule of land mammals, the septum nasi of cetaceans shows the following differences. The rostrum is disproportionally enlarged, which is particularly distinct at older stages. The connection of the septum with the original elements of the solum nasi which, in cetaceans, form the anterior side wall structures, is completely liquidated. The caudal edge of the septum falls off steeply and lies almost perpendicular with the longitudinal axis of the head. It is only in this place that the septum is still connected with the heavily reduced lateral parts of the nasal capsule, the tectum nasi and the lamina cribrosa. In addition, the cetacean septum lacks any perforations; there is no fenestra internasalis. A special form of the septum is found in *Physeter* (Klima et al. 1986; Klima 1987a), in which the upper edge of the septum emits

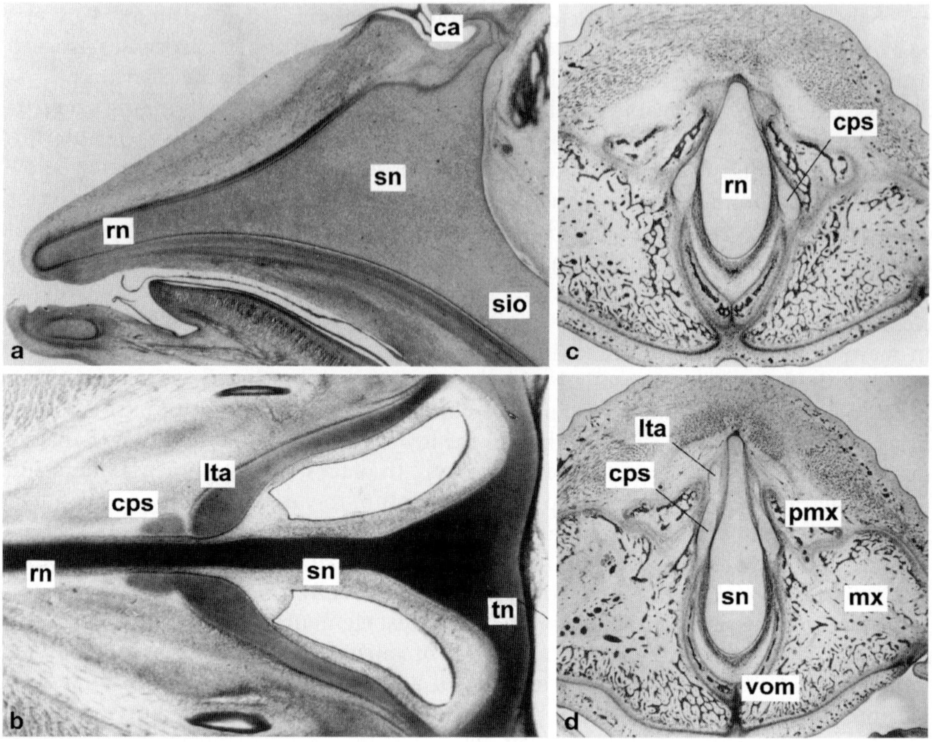

Fig. 38a-d. Septum nasi (*sn*). **a** *Stenella attenuata*, 85 mm TL, sagittal section. **b** *Monodon monoceros*, 137 mm TL, horizontal section. **c,d** *Stenella attenuata*, 115 mm TL, transverse sections from anterior (**c**) to posterior (**d**)

a branch that connects with the tectum nasi and forms with it a nasal roof cartilage (see Sect. 6.2.1).

The connection of the desmal elements with the cartilaginous septum is much more intense in cetaceans than in land mammals. The topographic relationships are basically altered by the far-reaching extension of the tectum nasi and the paries nasi, as well as by the altered position of the septum nasi, and above all, by the protrusion of the septum nasi as a single cartilaginous structure extending rostrally. As already seen in land mammals, the vomer embraces the lower edge of the septum, accompanying it along its entire length. In addition to that, the septum is also supported by the praemaxillare and the maxillare. The praemaxillare is intimate with the lateral surface of the septum, accompanying it all along its upper edge from the rostral tip towards the caudal end and turning laterally only in the place where the lamina transversalis anterior (see below) also separates from the septum. The maxillare lies close to the praemaxillare laterally, as well as to the vomer, accompanying the two desmal bones from the rostral tip caudally, and at the same time becoming intimate with the lateral surface of the septum close behind the cartilago paraseptalis (see below) and under the lamina transversalis anterior. All these structures, the cartilaginous septum as well as the three dermal bones mentioned, participate in the formation of the long rostrum, typical of the cetaceans, which no longer corresponds to the nasal skull but to the upper jaw.

6.1.2 Spina mesethmoidalis

At the upper caudal end of the septum nasi, before the lamina cribrosa, arises medially an unpaired, spiniform process, the spina mesethmoidalis. It develops in the place where the caudal edge of the tectum nasi joins the septum laterally in the form of an arch of limbus praecribrosus. This is a very variable structure, the origin of which is not quite clear. According to Henckel (1928), who denotes it as the protuberantia ethmoidalis posterior, it is a formation from the septum and the tectum nasi. Matthes (1921a), who examined this structure in detail, believed it to be a remainder of the original cartilaginous cranial tectum of vertebrates. His belief was corroborated by the finding of a paired anlage of the spina mesethmoidalis in *Manis* (Pholidota) by Starck (1941).

The spina mesethmoidalis is absent in Monotremata and Marsupialia. It is relatively frequent in the Placentalia, for example, in some Chiroptera (*Miniopterus*: Fawcett 1919; *Myotis*: Frick 1954), Dermoptera (*Cynocephalus*: Halbsguth 1973), Scandentia (*Tupaia*: Henckel 1928; Spatz 1964; Zeller 1983; 1985), Primates (*Propithecus*: Starck 1962; *Lemur*: Spatz 1964; *Saguinus*: Müller 1981; *Callithrix*: Simon 1981; *Galago*: Warich 1986), Xenarthra (*Bradypus*: Schneider 1955), Rodentia (*Cavia*: Rajtova 1972a, Zimmermann 1976, Dierbach 1985; *Ctenodactylus*: Schrenk 1989), Carnivora (*Canis*: Schliemann 1966), and Hyracoidea (*Procavia*: Lindahl 1948). On the other hand, the spina mesethmoidalis is absent in some of the Placentalia examined, for example, in Insectivora, Lagomorpha, Perissodactyla and Artiodactyla (De Beer 1937), many Primates (Maier 1983; Warich 1986) as well as in most Rodentia (Schrenk 1989).

It has been repeatedly emphasized in the literature that it is the Cetacea which possess a distinct spina mesethmoidalis. In Mysticeti this structure was already observed and illustrated by Eschricht (1849), and later by Kükenthal (1893) in *Balaenoptera*, yet it was De Burlet (1914a) who examined it likewise in *Balaenoptera* and denoted it as the spina mesethmoidalis. Honigmann (1915, 1917) found the spina mesethmoidalis in *Megaptera*. In our material, too, this structure is well developed in both these genera. It is robust and long, protruding steeply upwards from the level of the tectum nasi. However, Honigmann (1917) had already mentioned that its shape and size may vary. In Odontoceti, the spina mesethmoidalis has been described in *Phocoena* (De Burlet 1913a,b), *Lagenorhynchus* (De Burlet 1914b), *Monodon* (Eales 1950), *Physeter* (Klima et al. 1986), *Stenella* (Klima 1987a), and in the present paper, in *Delphinus*. The exact formation of the spina mesethmoidalis in *Globicephala* is not quite clear from the reconstruction illustrated by Schreiber (1916). Except for this latter uncertain case, the spina mesethmoidalis has been found in all Odontoceti examined so far. It is considerably smaller and especially shorter than in *Physeter* and in *Mysticeti*. As shown by our examinations of the developmental series of *Stenella*, *Delphinus* and *Phocoena*, the spina mesethmoidalis is rather erect in younger embryos; in older ones it declines posteriorly in the form of a caudal elongation of the upper edge of the septum nasi. In addition, its relative size decreases with the increasing age of the embryos.

The causes of the presence or absence of the spina mesethmoidalis are not clear. According to Schrenk (1989), it is developed only if it serves to fix the frontale or the falx cerebri. At the same time, however, Schrenk (1989) notes that these two functions can also be performed by other structures, such as the crista galli, with which the spina mesethmoidalis has often been mistaken. Also, the assumption of Spatz (1964) that the spina mesethmoidalis is suppressed by the robust development of the bulbi olfactorii in macrosmatic mammals (Insectivora), being able to develop only after a reduction of the sense of smell and the bulbi in microsmatic ones (*Tupaia* and *Lemur*), is unacceptable for several reasons, according to Zeller (1983). The assumption is indirectly refuted even by findings in whales, as the baleen whales possess a well-developed spina mesethmoidalis as well as a preserved bulbi olfactorii, whereas the toothed whales, with largely reduced bulbi, possess only a moderately developed spina mesethmoidalis.

Also, the repeatedly emphasized connection between the development of the spina mesethmoidalis and the adaptation to aquatic life habits may not necessarily hold true. It is true that, in addition to the Mysticeti and Odontoceti mentioned above, a well-developed spina mesethmoidalis is also found in Sirenia (*Halicore*: Freund 1908; Matthes 1921a, *Manatus*: Matthes 1912) as well as in both Pinnipedia examined (*Leptonychotes*: Fawcett 1918, *Mirounga*: Kummer and Neiss 1957). This tends to support the hypothesis formulated by Honigmann (1917), stating that the spina mesethmoidalis is found in all primordial craniums of typical aquatic mammals examined. However, Honigmann (1917) himself doubts this connection, stating that this structure would have to be considered an adaptation to aquatic life habits should it not disappear during subsequent development, which would naturally refute even this assumption. The topic was later discussed in detail several times (Matthes 1921a; De Beer 1937; Starck 1941) without any plausible explanation being offered, however. Therefore, Zeller (1983) is right in stating that it is quite uncertain whether and to what extent there is a connection between the presence of a spina mesethmoidalis and the aquatic life habits. Should there be such a connection, it might inhere in the fact that the generally somewhat elongated rostral part and the somewhat shortened frontal part of the skull of aquatic mammals are shifted into each other, obtaining a kind of a suture through the spina mesethmoidalis, which provides for a more firm connection of the two parts.

Certainly, the formation of the spina mesethmoidalis depends on species-specific differences, as indicated by the good examples of the two genera of Sirenia. *Halicore* possesses a spina mesethmoidalis that is unique among all mammals, reaching quite far caudally like an extremely long and arched cartilaginous spur under the dermal bones of the skull roof. In addition, another cartilaginous projection stems from the septum close behind the first one, denoted as the spina mesethmoidalis posterior by Matthes (1921a). On the other hand, the spina mesethmoidalis of *Manatus* is simple and short, resembling that of Odontoceti. These considerable differences do not tend to indicate any appreciable common mode in the otherwise equal adaptation of the two genera to the aquatic life habits.

The differences in the formation of the spina mesethmoidalis between baleen whales and toothed whales may be due to the basically different structure of their skulls. In all cetaceans the special functional requirements for the development of a long rostral part of the skull and its connection with the neurocranium lead to the typical overlapping of individual cranial bones. This phenomenon, denoted as "telescoping" by Miller (1923), shows a different tendency in the two groups (Fleischer 1976; Moore 1981). In the Mysticeti the posterior skull bones overlap the anterior ones, thus leaving enough room for the nasal skull. In the anterior wall of the neurocranium a fissure develops through which a long and robust spina mesethmoidalis can reach dorsally. In the Odontoceti the anterior skull bones overlap the posterior ones, leaving little room for the compressed nasal skull. Here only a short spina mesethmoidalis can develop, protruding posteriorly.

6.2 The Posterior Side Wall Structures

6.2.1 Tectum nasi

The tectum nasi is the most important part of the nasal roof. It is mostly situated praecerebrally, sending a smaller subcerebral part of the nasal roof caudally, the lamina cribrosa. The rostral end of the tectum is joined by the anterior closure of the nasal capsule, the cupula nasi anterior. On cross section, the nasal tectum shows the typical T-shaped construction, the middle vertical part being the septum and the upper lateral arms the tectum. The lateral arms overlap the two halves of the nasal capsule. Laterally, they are turned down and continue into the nasal side wall, the paries nasi. The cartilaginous tectum shows a marked connection with the desmal element, the nasale, which becomes intimate with the tectum from the dorsal end, accompanying it all along its length. This basic shape of the tectum nasi is found, without any appreciable alterations, in most land mammals (cf. De Beer 1937).

In cetaceans, the tectum nasi is considerably reduced. The heavily shortened tectum shows the praecerebral position typical of land mammals only in the baleen whales, *Balaenoptera* and *Megaptera* and in *Physeter*. It is situated between the cupula nasi anterior and the spina mesethmoidalis. Still more caudally it is joined by the slightly downturned lamina cribrosa. These structures are easy to interpret. Such a tectum nasi was observed and correctly interpreted in *Balaenoptera* as early as 1893 by Kükenthal. His concept was confirmed by further observations of De Burlet (1914a) in *Balaenoptera* and of Honigmann (1917) in *Megaptera*, and even the present findings have yielded the same results. The development of the desmal skull elements was examined by Schulte (1916) in *Balaenoptera*, and by Ridewood (1923) in *Balaenoptera* and *Megaptera*. Since the cartilaginous nasal roof is shifted rather far posteriorly in connection with the shifting of the nasal openings towards the vertex of the head, even the desmal elements accompanying the nasal roof are displaced together with it. At the same time, the plates become partly overlapping. Thus, the rostral part of the tectum is only covered with the nasale, its caudal part with the overlapping plates of the nasale and the frontale, and the major part of the nasale is externally turned.

Quite different relationships are found in Odontoceti (Fig. 39). In them the tectum nasi is compressed to form a short cartilaginous section which is shifted far posteriorly and turned into a predominately vertical position. It is no longer praecerebral but forms, together with the lamina cribrosa and the paries nasi, a common vertical plate which, on the one hand, closes the viscerocranium, and on the other, forms the anterior

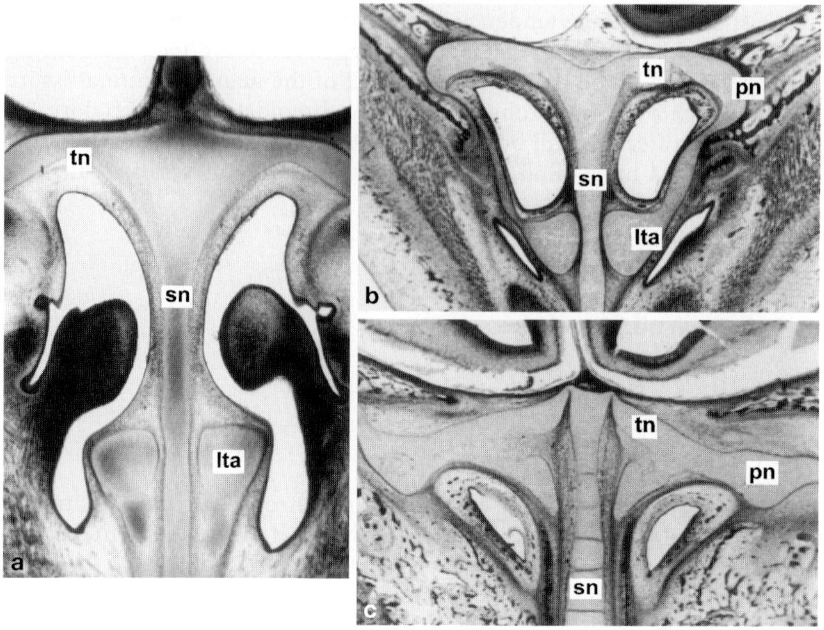

Fig. 39a-c. Tectum nasi (*tn*) in horizontal sections. **a** *Monodon monoceros*, 137 mm TL. **b** *Lagenorhynchus albirostris*, 107 mm TL. **c** *Phocoena phocoena*, 134 mm TL

wall of neurocranium. It is these structures that we have denoted, as a whole, the posterior side wall structures (Klima and Van Bree 1985; Klima 1987a). It is hardly possible to precisely define the boundary between the tectum and the remaining structures. With greatest probability, the material of the tectum nasi is situated partly medially in the short region between the free cupula nasi anterior and the spina mesethmoidalis, and partly laterally in the short bar that forms the downward connection with the lamina cribrosa and the lateral one with the paries nasi. It is precisely that region of the cartilaginous plate which is also accompanied by the equally heavily shrunk nasale. In this, the covering bones overlap similarly as in the baleen whales. The nasale is shifted externally so that the frontale lying under it partly overlaps the cartilaginous tectum.

Kükenthal (1893) wrote that in the Odontoceti examined by him, he was unable to find none of the rudiments of a nasal roof. De Burlet (1913a,b, 1914b) describes, in *Phocoena* and *Lagenorhynchus*, the tectum nasi as consisting of only the anterior, free part that is not accompanied by the nasale and that arches around the nasal aperture; in our concept, this corresponds to the cupula nasi anterior (Klima and Van Bree 1985; Klima 1987a). The interpretation of De Burlet (1913a,b, 1914b) has also been advocated by De Beer (1937) who characterizes the results obtained from *Globicephala*, and adopted in the same sense by Schreiber (1916), as follows: "The roof is formed of a small bar connecting the lamina orbitonasalis with the nasal septum, and of a pair of isolated cartilages situated a little farther forward." Similarly, Eales (1950) denotes as the tectum nasi in *Monodon* only the anterior part of this structure. Our later examinations in *Monodon* (Klima and Van Bree 1985), however, have revealed a

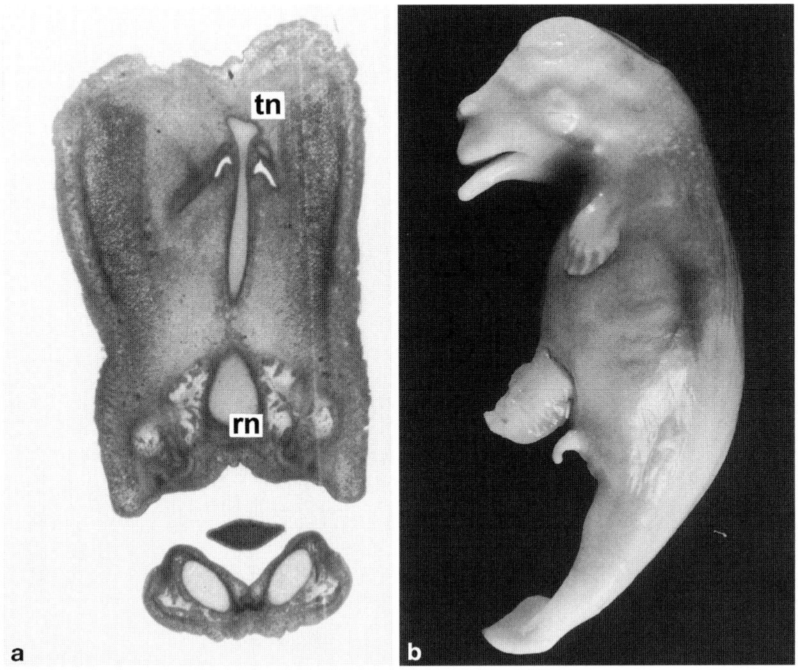

Fig. 40a-b. Tectum nasi (*tn*) in *Physeter catodon*, 97 mm TL. **a** Transverse section of the forehead. **b** Appearance of the embryo; note the shape of the head

relatively well developed tectum reaching rather far caudally and ventrally, and forming, with the septum, a massive construction, T-shaped in cross section (Fig. 39a). Similar relationships, with only slight alterations, are found in all other Odontoceti examined, i.e., in *Stenella, Delphinus, Lagenorhynchus* and *Phocoena.* An exception is provided by the tectum nasi of *Physeter* (Klima et al. 1986; Klima 1987a), which assumes a different function, reaching together with a branch of the septum nasi, far anteriorly in the form of a "nasal roof cartilage" (Figs. 40, 41).

6.2.2
Lamina cribrosa

In the opinion of most authors (Gaupp 1900, 1905; Voit 1909; De Beer 1937; Slabý 1960; Starck 1967; Kuhn 1971; etc.), the lamina cribrosa is a neomorph structure in mammals, developing only in connection with the development of the sense of smell. It is a continuation of the tectum nasi in the caudal direction and is predominately situated subcerebrally. Due to this position, it is functionally the floor of the neurocranium. The slightly concave plate forms the supporting plane for the bulbus olfactorius, and at the same time, its mostly numerous apertures, the foramina olfactoria, enable the passage of the fila olfactoria from the olfactory mucosa of the nasal cavity to the bulbus olfactorius of the brain. Of course, the absence of lamina cribrosa in some mammals,

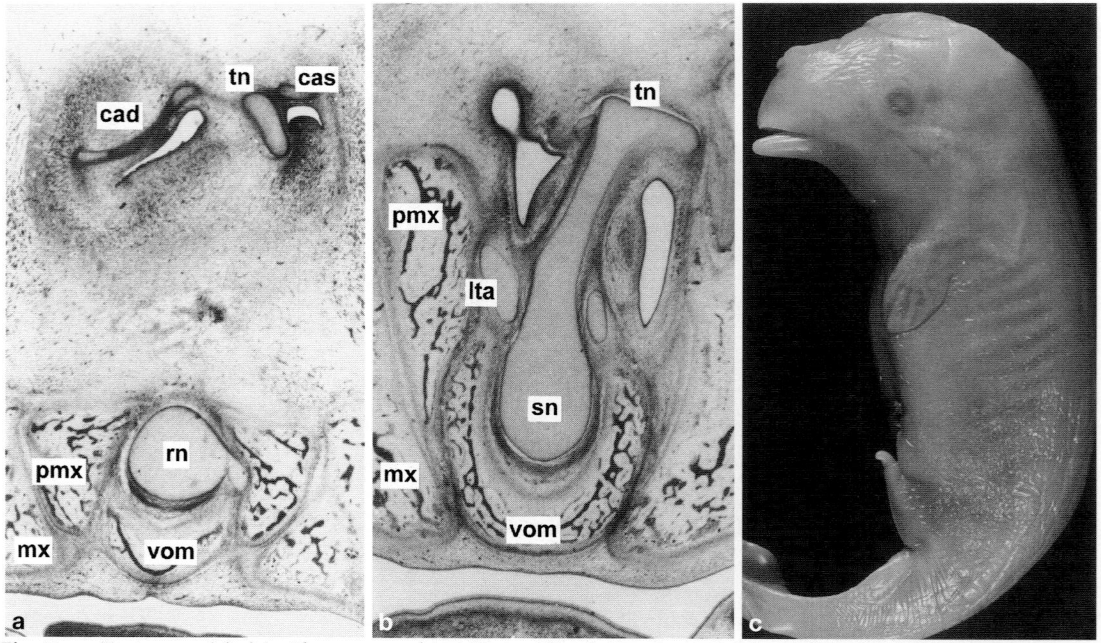

Fig. 41a-c. Tectum nasi (*tn*) in *Physeter catodon*, 150 mm TL. **a,b** Transverse sections of the forehead from anterior to posterior. **c** Appearance of the embryo; note the shape of the head e.g., in *Ornithorhynchus* (Monotremata), is not a primitive reptilian characteristic, as was previously thought (De Beer 1937; Vandebroek 1964), but rather a secondary condition, as ascertained by Kuhn (1971) in a comparison with *Tachyglossus* (Monotremata). Recently, the development of the lamina cribrosa was studied in detail by Zeller (1988) who has formulated his results as follows: "As in many other water-adapted mammals, the olfactory system of *Ornithorhynchus* is reduced. This suggests that the rest of the lamina cribrosa of *Ornithorhynchus* is secondarily reduced. The common ancestor of *Ornithorhynchus* and *Tachyglossidae* most probably possessed a lamina cribrosa developed only once in the phylogeny of mammals. Its lack in adult *Ornithorhynchus* is not a reptilian character."

In mammals adapted to aquatic life habits, the sense of smell has generally lost its importance. This is particularly true of the holaquatic Cetacea, the organs of smell as well as the olfactory brain of which are heavily reduced, apparently in favor of the sense of hearing, a much more effective system serving both orientation and communication under water. Nevertheless, their olfactory system has not been reduced completely. At the same time, the reduction is less advanced in the baleen whales than in the toothed ones. Oelschläger (1992) writes: "Adult baleen whales retain a reduced but certainly functional main olfactory system ... They have inconspicuous olfactory bulbus and very slender olfactory tracts which resembles the situation in pinnipeds and humans." He goes on to add:"... in adult odontocetes the rostral olfactory system is totally reduced."

In contrast to the adults, the embryos of Odontoceti still possess a bulbus olfactorius as well as the fila olfactoria which pass through the lamina cribrosa and connect the olfactory epithelium of the nasal mucosa with the bulbus. The anlage of the bulbus olfactorius disintegrates already at the early stages of fetal development. The fila olfactoria receive an additional, non-olfactory component, the terminalis system. While the olfactory system disintegrates, the terminalis system remains preserved. It stretches from cell nests resembling ganglia, lying near the septal nasal mucosa, over the nerve filaments through the lamina cribrosa to the ganglion terminale inside the primitive meninx, and thence over several bundles of the nervus terminalis to the telencephalon (Buhl and Oelschläger 1986, 1988; Oelschläger et al. 1987; Oelschläger 1989, 1992). In the Odontoceti the terminalis system shows a progressive development, a unique phenomena among mammals. Oelschläger (1992) writes: "While the olfactory complex is totally reduced (adult toothed whales are anosmatic), the number of terminalis neurons in dolphins attains a maximum within mammalia." It is still unknown whether even adult baleen whales possess a similarly highly developed terminalis system. Oelschläger (1992) thinks that "... it seems plausible that they will not show such a hypertrophy of the terminalis system as toothed whales." The still not quite clear functional significance of that system has been the topic of recent discussions (cf. Sinclair 1951, 1966; Buhl and Oelschläger 1986; Ridgeway et al. 1987; Oelschläger et al 1987; Oelschläger 1989, 1992).

The embryonic lamina cribrosa of cetaceans develops as a mesenchymatous and later cartilaginous partition between the peripheral and central parts of the olfactory or the terminal system. It remains open in places, thus forming passages for neural connections. In contrast to the numerous apertures in the lamina cribrosa of land mammals, that of the cetaceans shows apertures in just a few places (Fig. 42).

The lamina cribrosa of embryonic baleen whales, *Balaenoptera* and *Megaptera*, forms only a narrow margin and a single large opening, the foramen olfactorium (see

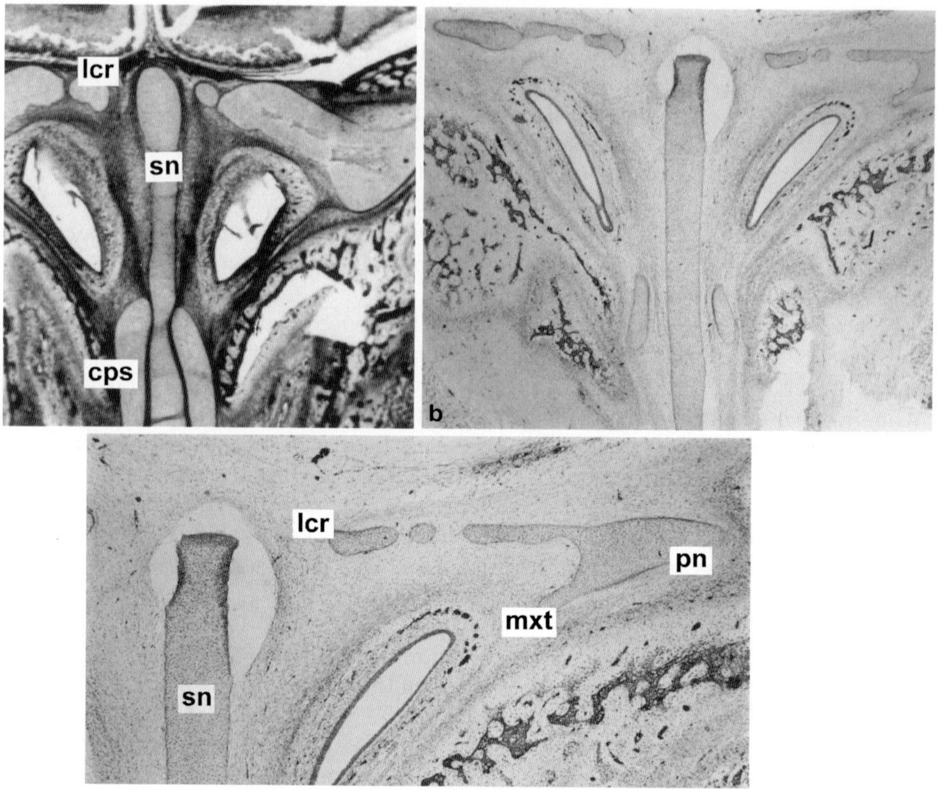

Fig. 42a-c. Lamina cribrosa (*lcr*). **a** *Lagenorhynchus albirostris*, 107 mm TL, horizontal section. **b,c** *Stenella attenuata*, 89 mm TL, horizontal section in overall view (**b**) and in detail (**c**)

Fig. 45). Medially, this opening is delimited by the upper edge of the septum nasi. Laterally the edge of the lamina cribrosa continues into the paries nasi; rostrally, into the tectum nasi; and caudally, into the upper edge of the cupula nasi posterior. In the direction from rostral to caudal, the surface of the lamina cribrosa drops down at an angle of about 45°. In this way it is partly subcerebral and partly praecerebral. From above, i.e., from the neurocranium, the bulbus olfactorius inserts into the foramen olfactorium its ganglion terminale. From below, i.e., from the nasal capsule, the foramen olfactorium is entered by the bundles of the fila olfactoria and the fila terminalia (Oelschläger 1989, 1992).

Already the earlier papers on *Balaenoptera* (Kükenthal 1893; De Burlet 1914a) and *Megaptera* (Honigmann 1915, 1917) describe and illustrate the position and shape of the lamina cribrosa with its single foramen olfactorium, in a similar manner as the present contribution. These papers also report on the topographic connections of the lamina cribrosa with the "lobus and nervus olfactorius," even if without dividing it into the olfactory and the terminal systems. De Burlet (1914a) writes the following about the lamina cribrosa of *Balaenoptera* (105 mm): "The foramina cribrosa are essentially substituted, on either side, by a single large opening (the fenestra cribrosa).

Suggestions of a division are present in the form of cartilaginous bridges in just a few places at the distal and lateral edges of this opening. It is probable that the development of the cartilaginous bridges in the lamina cribrosa have not yet reached the peak of development in the embryo examined." Nothing is known about the subsequent course of the morphogeny of the lamina cribrosa in baleen whales. However, it has long been known that the lamina cribrosa of adult baleen whales possesses up to three rows of smaller passages for the fila olfactoria. That distinct nervi olfactorii still occur in adult baleen whales is "... an old fact that has been certainly demonstrated by Eschricht's (1869) illustration of the brain of *Megaptera*" (Kükenthal 1893).

In embryonic toothed whales the lamina cribrosa is almost vertical. In this way it assumes secondarily a praecerebral position and participates in the formation of the anterior wall of the neurocranium. When fully developed, it possesses several apertures, the largest one being situated medially (see Figs. 6–8). It is fissure-like and lies between the septum nasi medially and the longitudinal edge of the lamina cribrosa laterally. Dorsally it begins behind the plate connection of the tectum nasi with the septum. Ventrally it ends in the region of the cupula nasi posterior. In cases where the cupula nasi posterior is continuously connected with the septum, the opening finds here its lower delimitation (*Phocoena*, *Physeter*). In cases where these structures are not connected but merely approximated, the fissure-like opening is open below (*Stenella*, *Delphinus*, *Lagenorhynchus*, *Monodon*). Laterally there is yet another, or several additional, openings. They are evidently rudimentary in character, even if they temporarily serve as passages for nerves. Their number and size depend upon the developmental degree of the nasal skull, and besides, they show considerable variation. Moreover, the most lateral openings are largely closed by a reversed fold of the maxilloturbinale from the rostral direction.

Data on the lamina cribrosa of Odontoceti are scarcely found in the literature. Kükenthal (1893) mentions, in his paper, the embryonic lamina cribrosa of Odontoceti in a very general manner, without going into particulars. Also, his illustrations show no suitable cross section through the lamina cribrosa. De Burlet (1914b) describes the lamina cribrosa of *Lagenorhynchus* as a vertical plate, which forms the anterior closure of the cerebral capsule. As regards the lamina cribrosa of *Phocoena*, De Burlet (1913a) writes: "Since the nervus olfactorius is completely absent in adult *Phocoena*, no extensive perforation of the lamina cribrosa can even be expected to occur in the embryo. Nevertheless, there are a few openings, a larger, lateral one on either side, and two smaller ones." More detailed data on the lamina cribrosa of Odontoceti are only found in the papers by Klima and Van Bree (1985) and Klima et al. (1986), as well as in part in the neuroanatomical publications of Buhl and Oelschläger (1986), Oelschläger et al. (1988) and Oelschläger (1992).

6.2.3 Paries nasi

The side wall of the nasal capsule is thought to develop, during ontogeny, from three paired cartilaginous centers: an anterior cartilago parietotectalis, a medial cartilago paranasalis and a posterior cartilago antorbitalis, which unite to form a continuous cartilaginous plate, the paries nasi (Nordenbos 1905; Terry 1917; Reinbach 1953; Zeller 1983). However, this developmental mode cannot be demonstrated in all mammals.

Already at the earliest stages, the paries nasi can be formed from a single anlage (e.g., in *Myotis*, Chiroptera: Frick 1954).

Dorsally, the paries nasi is a continuation of the tectum nasi, there being no distinct division line. The plate of the paries nasi bends down laterally, its free ventral edge ending in an inner furl, the maxilloturbinale. The bent plate of the paries nasi is rostrally delimited by the cupula nasi anterior, caudally by the cupula nasi posterior. The external surface of the paries is divided into a pars anterior, a pars lateralis and a pars posterior (Voit 1909), defined by two vertical transverse depressions, the sulcus lateralis anterior and the sulcus lateralis posterior. The anterior sulcus corresponds, on the inner surface, to the projection of the crista semicircularis, the posterior sulcus to that of the ethmoturbinale I. The inner surface of the paries nasi is rugged and forms the proper inner space of the nasal capsule. The outer surface is rather flat, without any marked projections. The external shape of the paries nasi is mainly determined by its middle section which mostly bears a lateral vault, the prominentia lateralis. In the broadest place, at the passage to the tectum nasi, the commissura orbitonasalis or its rudiment usually projects.

In all Cetacea the paries nasi is heavily reduced and particularly shortened. At the same time, the delimitation of this structure remains the same as in other mammals, i.e., it lies between the tectum nasi, the maxilloturbinale, the cupula nasi anterior and the cupula nasi posterior. It possesses a commissura orbitonasalis, forming the connection to the ala orbitalis. In Mysticeti, the longitudinal axis of the paries nasi is inclined from anterior dorsal to posterior ventral at an angle of about 45°. Its outer wall is very convex, pointing laterally. In Odontoceti the longitudinal axis of the paries nasi is vertical. Its wall is heavily flattened and points caudally.

While the wall of the paries nasi of Odontoceti bears no other projections besides the commissura orbitonasalis, in the Mysticeti examined, two small cartilaginous processes have been ascertained. The upper one lies in the region of the upper paries in front of the protrusion of the commissura orbitonasalis. The lower process lies in the region of the ventral edge of the paries in front of the cupula nasi posterior. The homology as well as the function of these cartilaginous structures are uncertain.

The upper projection has been found in both genera of Mysticeti examined. In *Balaenoptera* (see Figs. 19, 20) it is situated rather high and shifted slightly rostrally of the commissura orbitonasalis. It is connected with the external wall of the paries through a narrow shaft and points anteriorly like a finger. De Burlet (1914a) denotes it as the processus paranasalis. De Beer (1937) considers it to be a "parethmoidal cartilage." In *Megaptera* (see Figs. 23, 24) this cartilaginous processus is approximately the same in shape and size, but is shifted somewhat more caudally and ventrally. Nevertheless, it is unequivocally situated in front of the commissura orbitonasalis. In the reconstruction by Honigmann (1917) the processus paranasalis is situated still somewhat deeper and its connection with the commissura orbitonasalis is not clearly indicated. For this reason, De Beer (1937) considers it to be not the upper but the lower projection, and he denotes it as the processus maxillaris anterior. This is certainly wrong, as indicated by a comparison with our reconstruction.

It is questionable whether this upper projection in *Balaenoptera* and *Megaptera* actually corresponds with the processus paranasalis. Nevertheless, most of the circumstances support this interpretation. In a comparable site, a processus paranasalis has been found in several mammals, e.g., in *Dasypus*, Xenarthra (Reinbach 1952), *Pongo*, Primates (Starck 1960) and, above all, as a marked structure in *Homo*

(Von Michalkovics 1899; Frets 1914; Macklin 1914, 1921; Hauschild 1937; Reinbach 1963; Bersch and Reinbach 1970). It has about the same comparable shape and size, it points rostrally and is continuously connected with the paries nasi. From the exterior it is embraced by the maxillare. According to Reinbach (1963) it shows a close connection with the ductus lacrimonasalis, embracing it, in mammals, sometimes dorsally and sometimes laterally. However, I have been unable to ascertain this connection in the Mysticeti examined, as they lack a ductus lacrimonasalis.

According to Reinbach (1963), the parethmoidal cartilage, likewise occurring in humans in this region, is situated more dorsally and is significantly smaller. It is mostly separated from the nasal wall so that it develops not in the form of a projection but as an isolated cartilago paraethmoidalis. This cartilage is embraced by the lacrimale. In Mysticeti none of the properties mentioned hold true for the upper projection and, hence, it cannot be considered to be the cartilago paraethmoidalis.

Yet another possibility could be taken into consideration. In *Oryctolagus*, Lagomorpha (Frick and Heckmann 1955) there is, in an approximately comparable site in the paries nasi, close before the commissura orbitonasalis, a similarly shaped cartilaginous projection, the trochlea for the musculus obliquus superior of the bulbus oculi. In the Mysticeti examined, however, there is no connection with the ocular musculature. The projection is rather far away from the eye and is deeply inserted between the paries nasi and the maxillare. Thus, in this case the trochlea cannot be involved.

The lower projection has only been ascertained in *Balaenoptera* (see Fig. 20). De Burlet (1914a) describes it as three small cartilaginous bars situated one above the other approximately in the direction of the nasal axis. In our material, this triple form of the projection is not discernible. From the outer wall of the paries nasi rises a slightly irregularly shaped ledge from which a finger-like projection protrudes rostrally like an arch. This projection is absent in *Megaptera*. The upper projection, the processus paranasalis, illustrated in the *Megaptera* reconstruction by Honigmann (1917), however, is situated so deeply that De Beer (1937) considered it to be the lower projection and denoted it as the processus maxillaris.

The interpretation of the lower projection remains uncertain. De Burlet (1914a) dares not name this projection at all, but mentions the possibility that it could be the processus maxillaris posterior, which has been found in this place in some mammals. The author gives examples of projections found on the cartilaginous skulls of the pig, cattle and sheep, and described by Spöndli (1846) and Decker (1883). De Beer (1937) adopted, for his monograph, only that part of the *Balaenoptera* reconstruction by De Burlet (1914a) which does not illustrate this structure, and that is why it does not appear in his considerations at all. According to Reinbach (1963), the processus maxillaris posterior, described in *Homo* by Macklin (1914) and Hauschild (1937), is a denomination justified by nothing, as apparently the cartilago paraethmoidalis is the case. The literature so far contains probably several cases of mistakes for other structures. According to Zeller (1983), a true processus maxillaris posterior has been demonstrated in only a few mammals, which may be due to the fact that it occurs only during a short period of ontogeny. In *Dasypus* and *Zaedyus*, Reinbach (1952, 1953) found, at the posterior lower part of the nasal capsule, similar projections, which he considers to be remains of the palatoquadratum, and which he describes as the cartilago antorbitalis. The homologies are questionable and our material does not enable us to contribute to clarify them. In spite of all reservations, the interpretation of the lower projection in *Balaenoptera*, already anticipated by De Burlet (1914a), as the processus maxillaris posterior seems to me to be the most probable.

6.2.4 Commissura orbitonasalis

The regio ethmoidalis and the regio orbitotemporalis of the embryonic chondrocranium are connected, in the lateral region, by a paired cartilaginous bridge, the commissura orbitonasalis. This commissure originates from the lateral plate of the nasal capsule in the place where the tectum and the paries nasi contact (Kuhn 1971), and it reaches to the anterior edge of the ala orbitalis. Together with the rostral part of the ala orbitalis, this cartilage should be homologous with the reptilian cartilago spheneth-moidalis (Reinbach 1952). Previously, it was denoted as the commissura sphenoethmoidalis (cf. Gaupp 1900, De Beer 1937). Besides the unpaired septum nasi, the paired commissura orbitonasalis, if present forms the only additional connection between the nasal capsule and the rest of the chondrocranium. Hence, it is assumed that it plays a mechanical role in strengthening the entire chondrocranium.

In Monotremata (*Ornithorhynchus*: De Beer and Fell 1936; *Tachyglossus*: Kuhn 1971) the commissura orbitonasalis is very much enlarged, and it even participates in the formation of the tectum neurocranii. Also, all Marsupialia examined (*Perameles*: Esdaile 1916; *Dasyurus*: Fawcett 1917; *Didelphis*: Toeplitz 1920; *Caluromys*: Denison and Terry 1921; *Wallabia*: Müller 1986; *Monodelphis*: Maier 1987) possess a robust commissura orbitonasalis which forms a continuation between the nasal capsule and the neurocranium. According to Maier (1987), this provides for the functionally necessary stiffening of the rostral skull region to meet the special requirements of the neonates and pouch young of Marsupialia.

While the commissura orbitonasalis is present in most Placentalia, it is only moderately to weekly developed. It has been ascertained, e.g., in some Insectivora (*Talpa*: Fischer 1901; *Erinaceus*: Fawcett 1918), Chiroptera (*Miniopterus*: Fawcett 1919; *Myotis*: Frick 1954), Dermoptera (*Cynocephalus*: Henckel 1929), Scandentia (*Tupaia*: Henckel 1928; Spatz 1964; Zeller 1983, 1987), Xenarthra (*Bradypus*: De Burlet 1927, Schneider 1955; *Dasypus*: Fawcett 1918; Reinbach 1952), Lagomorpha (*Lepus*: Voit 1909; De Beer and Woodger 1930), Sirenia (*Halicore*: Matthes 1921a); Perissodactyla (*Equus*: Muggia 1931), and Artiodactyla (*Sus*: Mead 1909; *Bos*: Fawcett 1917, 1918; De Beer 1937). A developed commissura orbitonasalis is also present in all examined Carnivora (cf. Schliemann 1987; Timm et al. 1989) and Rodentia (cf. Schrenk 1989), except for *Bathyergus* (Rodentia) in which the commissure remains separated from the nasal capsule.

Most primates also possess a commissura orbitonasalis, (e.g., *Macaca*, *Semnopithecus*: Fischer 1903; *Nycticebus*: Henckel 1927; *Papio*: Reinhard 1958; *Pan*: Starck 1959; *Saguinus*: Müller 1981; *Galago*: Warich 1986; *Daubentonia*: Maier 1991). However, their commissura orbitonasalis, instead of being mostly vertical as in most mammals, is almost horizontal. According to Kuhn (1971), this change in positions is due to the intense development of the brain in the region of the fossa cranii anterior. In *Homo*, the commissura orbitonasalis has been transformed into a particularly broad, horizontal cartilaginous plate (Fawcett 1910; Macklin 1914, 1921; Reinbach 1963; Bersch and Reinbach 1970), obviously as a result of the significant increase in the size of the neurocranium and the simultaneous shift of the nasal skull to the subcerebral position. On the contrary, the commissura orbitonasalis of *Propithecus* (Starck 1962) is considerably reduced, and it is virtually absent in *Tarsius* (Fischer 1905; Henckel 1927, 1928; Wünsch 1975). The loss of the commissura orbitonasalis in *Tarsius* is explained by Wünsch (1975) as resulting from the development of oversized eyes. Due to this, the commissure, which otherwise in other primates with less developed eyes, lies dorsal to the medial wall of the orbita. The mechanical stiffening of the nasal capsule against the neurocranium is provided by the covering dermal bones of the facial region. They embrace the nasal capsule from oral and dorsolateral positions. As a matter of fact, the unusual development of the sense of vision in *Tarsius* leads to extensive changes in the embryonic skull (Wünsch1975; Starck 1982, 1984).

The Mysticeti *Balaenoptera* (De Burlet 1914a; present report) and *Megaptera* (Honigmann 1915, 1917; present reports) possess a commissura orbitonasalis which is comparable with that of most other Placentalia (see Figs. 19–21, 23–25). It originates above a lateral position on the paries nasi about the place in which the tectum nasi and the lateral edge of the lamina cribrosa join the paries. It first passes in a short, upwardly lateral arch and then assumes a caudal direction, reaching the ala orbitalis. It consists of a robust cartilaginous strut round in cross section. Following the line of this strut beyond its limits, it becomes apparent that it is a component of a thorough, longitudinal cartilaginous strip passing from a rostral direction over the cartilago paraseptalis, the lamina transversalis anterior and the upper part of the paries nasi into the commissura orbitonasalis and reaches far caudally over the upper edge of the ala orbitalis. This system of cartilaginous struts, together with the covering dermal bones, the praemaxillare and the maxillare, forms a firm lateral anchoring of the protruding embryonic rostrum at the neurocranium.

In all Odontoceti examined, save for a single exception, the commissura orbitonasalis bears only a rudimentary character or is virtually absent. It can be observed in the developmental series of *Stenella* and *Delphinus* that a thorough commissura orbitonasalis is present only at the younger stages. At the older ones, it is being reduced. In this, the reduction need not proceed symmetrically at all. In one and the same embryo the commissure can still be present on one side and be already reduced on the other side of the head. Even the time at which the reduction takes place may vary and may cover several developmental stages. The origin of the commissura orbitonasalis is shifted down to the lateral lower edge of the posterior side wall structures of the cetacean skull, far away from the position typical of the remaining Placentalia. At the same time, one must admit that even in Odontoceti the commissura orbitonasalis originates in the region where the paries edges with the tectum. Yet these structures are heavily deformed and far displaced, as seen above. This also causes the shift of the commissura orbitonasalis. If present, the commissura consists of only a narrow cartilaginous strip of obviously unimportant mechanical action. A reduced commissure remains preserved in the form of a small projection on the posterior side wall structures. Even the examined younger embryos of *Lagenorhynchus* possess the commissura orbitonasalis (De Burlet 1914b), whereas in the older ones (Klima and Van Bree 1985; present report) it is reduced to only small projections. In the only embryo of *Monodon* examined, neither Eales (1950) nor Klima and Van Bree (1985) found a preserved commissura orbitonasalis. However, this embryo belongs to the size category in which, in *Stenella* and *Delphinus*, the commissura orbitonasalis has already been reduced. Therefore, it could be well possible that a fully developed commissure does exist at the younger stages of *Monodon*. The same holds true for the only embryonic *Globicephala* examined (Schreiber 1916). In contrast, a commissura orbitonasalis has never been found in any embryo of *Phocoena*, not even in the form of a rudimentary projection, although it is in this species that many embryos have been examined, from the smallest up the largest ones (De Burlet 1913a,b; Klima and Van Bree 1985).

The reduction of the commissura orbitonasalis in Odontoceti is due to the following circumstances. On the one hand, the tectum as well as the paries nasi are heavily reshaped. Both these structures are shifted far posteriorly and change their position from originally horizontal to vertical. Due to this reshaping and changed position, even the commissura orbitonasalis is pushed far in a posterior ventral direction. On the other hand, the rostral part of the skull of Odontoceti is not only elongated, but in contrast to the baleen whales, it is also unusually high in its caudal section. The vertical posterior edge of the septum as well as the whole plate of the posterior side wall structures become intimate with the anterior wall of the likewise erect neurocranium. The large supporting surface provides a firm connection of the two skull parts, rendering the commissura orbitonasalis unnecessary. At later developmental stages, an additional anchoring of the rostral skull at the neurocranium is provided by the overlapping dermal bones, the praemaxillare and the maxillare, as in the baleen whales. Thus, the Odontoceti is the sole mammalian group in which a complete reduction of the commissura orbitonasalis occurs.

An exception to this rule is provided by the embryonic skull of *Physeter*. Not only because the commissura orbitonasalis is wholly retained: in this case it is even doubled (Klima et al. 1986). One can distinguish here a lower and an upper branch of the commissura orbitonasalis (Klima 1987a). The lower branch originates on the lower edge of the posterior side wall structures and points caudally, about the place in which a rudimentary commissura orbitonasalis is present in other Odontoceti. The upper branch originates on the upper edge of the posterior side wall structures, lateral of the spina mesethmoidalis. Both these origins lie on the long vertical connecting line between the tectum and the paries nasi, that is, on the line where the commissura orbitonasalis normally originates in other mammals as well. Both branches are robust and they join caudally the upwardly elongated ala orbitalis. They are fully developed in the smallest as well as the biggest embryos examined (see Figs. 16–17).

Beyond doubt, the doubling of the commissura orbitonasalis in *Physeter* is connected with the development of the highly specialized, enormous anterior part of the head, a unique formation among all cetaceans. During embryogeny, two cartilaginous struts are responsible for the development and formation of this “soft” and mostly skeleton-less anterior part of the head. The lower strut corresponds to the rostrum nasi of other cetaceans, and it becomes a component of upper yaw. The upper strut is the product of fusion of the tectum with the septum nasi. It passes through the anterior part of the head obliquely towards the upper anterior and accompanies, as the “nasal roof cartilage,” the left nasal passage. The two branches of the commissura orbitonasalis can be considered to be a continuation of these two cartilaginous struts in the caudal direction, since they originate in almost precisely the same height on either side. In this way a system of thorough cartilaginous struts develops, which fix the anterior head to the neurocranium at two superimposed levels. Certainly, this supporting system is a special adaptation and a unique phenomenon among mammals. The dermal bones adopt the shape of the cartilaginous model and accompany the lower rostral cartilaginous strut as well as the whole region in which lie the two branches of the commissura orbitonasalis. The upper cartilaginous strut, the "nasal roof cartilage," being a new formation, is not accompanied by dermal bones.

6.2.5
Cupula nasi anterior

The cupula nasi anterior forms the anterior closure of the embryonic nasal capsule. On either side of the septum nasi, it forms an approximately ring-like strut of the external nostril, the apertura nasi externa. This is not a unitary structure. In most cases it is considered to be a complex of several elements which are very variable and difficult to homologize (cf. Gaupp 1908; De Beer 1937; Reinbach 1952; Frick 1954; Starck 1967; Kuhn 1971; Maier 1981; Schrenk 1989).

The dorsally situated cartilago cupularis, or the cupula nasi anterior in the narrower sense, is the largest and most constant element of this complex. Being a continuation of the tectum nasi, it is bent down from a dorsal direction over the anterior edge of the septum nasi on both sides, thus forming a T-shaped closure of the nasal capsule rostrally. This element also mostly gives rise to the later cartilago alaris. The lower edge of the cartilago cupularis may be formed into two variable projections, viz., an anterior processus cupularis and a processus alaris posterior. Besides, there is, in this region, a close relation to the processus alaris superior, an independent element of the nasal skeleton. Still more laterally and ventrally, the cartilaginous ring continues, in a very inconstant manner, over the crista marginalis and the lamina transversalis, until it attains, over the processus lateralis ventralis, the lower edge of the septum nasi, thus completing the ring. The two latter structures, however, distinctly tend to separate and cease to serve as a support of the external nasal aperture. This is also the case in Cetacea (see Sects. 6.3.1, 6.3.2).

A complete cupula nasi anterior, consisting of several elements, is thought to represent a primitive characteristic in mammals (Sturm 1937). In fact, a rather complete cupula nasi anterior has been observed in only a few mammals examined, for example in *Didelphis* (Toeplitz 1920), *Talpa* (Fischer 1901), *Myotis* (Frick 1954), *Rousettus* (Starck 1943) and *Tupaia* (Zeller 1983). On the other hand, in many mammalian groups, a far-reaching simplification of these structures has been observed, e.g., in some primates (Maier 1981), Carnivora (Kummer and Neiss 1957; Schliemann 1966, 1987, 1991; Köhncke 1980, 1985; Timm et al. 1989), Lagomorpha (Voit 1909; Eloff 1950; Frick and Heckmann 1955) and Rodentia (Eloff 1948; Zimmermann 1976; Dierbach 1985; Hauck 1987; Schrenck 1989). Often the ventrally situated elements of the cupula nasi anterior are heavily reduced, and at the same time, the dorsally situated cartilago cupularis is increased in size and separated, e.g., as in some primates (Maier 1981, 1991, 1993) or in Artiodactyla (Sturm 1937). Functionally, this may lead to a greater mobility of the whole rostrum (Maier 1981), and at the same time, favor the tactile function of the external nasal parts (Halata 1975; Klauer 1976; Menzel 1979; Maier 1981). Beyond doubt, the cupula nasi anterior is the most variable part of the whole embryonic nasal capsule, depending on the different requirements of the functional adaptations. In aquatic mammals, the elements of the cupula nasi anterior are conspicuously reduced and transformed as a result of the heavy reduction of their sense of smell, as already observed by some authors (Sirenia: Matthes 1921a,b,

1923; Pinnipedia: Kummer and Neiss 1957). This is particularly true of Cetacea (De Burlet 1913a,b, 1914a,b, 1916; Honigmann 1915, 1917; De Beer 1937; Eales 1950; Klima and Van Bree 1985; Klima et al. 1986; Klima 1987a). According to the different degrees of this reduction or transformation, three different types of the cupula nasi anterior can be distinguished among the cetaceans.

The first type occurs in Mysticeti. The cupula nasi anterior rises towards the upper anterior region like a massive continuation of the tectum nasi. However, only the dorsal part of the complex has been preserved, whereas the lateral or ventral parts have been completely reduced. The dorsal part, which may correspond to the cartilago cupularis, produces rostrolaterally two (*Megaptera*) or three (*Balaenoptera*) projections (see Figs. 19, 23). There are practically no clues by which to homologize them. Nevertheless, Honigmann (1917) denotes, in *Megaptera*, only the medial projection as the cartilago cupularis, and he believes that the lateral projection could be the processus alaris superior. De Beer (1937), who has adopted Honigmann's (1917) reconstruction of the skull of *Megaptera* for his monograph, denotes the whole cupula nasi anterior as the processus alaris superior, which is certainly wrong. In *Balaenoptera*, De Burlet (1914a) only mentions a cartilago cupularis, however, with his illustration of this reconstruction being rather incomplete in the very region of the cupula nasi anterior. Our reconstruction in *Balaenoptera* clearly shows a tripartition, yet it seems to me too uncertain to interpret the individual elements as independent structures. For this reason, I have denoted the whole complex as the cupula nasi anterior. This cartilaginous complex embraces the nasal aperture only incompletely, in a semi-circle on its dorsal, i.e., posterior edge.

The second type occurs in Odontoceti. The cartilaginous structures are less massive but mostly encompass the whole nasal aperture. At the same time, even here the primary ventral connection to the septum nasi through a processus lateralis ventralis has been lost (see Sect. 6.3.2). In case the cupula nasi anterior is connected with the septum nasi not only dorsally (i.e., posteriorly in the Odontoceti), but also ventrally (i.e., anteriorly), thus forming a complete ring around the nasal aperture, this takes place secondarily through a very long and narrow processus growing ventrally (i.e., anteriorly) out of the cartilago cupularis, which is situated dorsally (i.e., posteriorly). In cetaceans, like in most Eutheria, the anlagen of the structures of the cupula nasi anterior appear relatively late in ontogeny, not until most other parts of the nasal capsule have already developed. Therefore, there are considerable differences in size and shape of the cupula nasi anterior between different developmental stages. The cupula nasi anterior is fully developed only in the older stages, and only then would it be possible to exactly compare the individual species, which, however, the examined material does not permit. With some reservation, the following characteristics of the cupula nasi anterior can be established at least for four odontocete genera. In *Lagenorhynchus* it forms a thorough ring. In *Delphinus* and *Stenella* it reaches far anteriorly in the form of a narrow arched process, ending free. In *Phocoena* it mostly loses connection to the tectum nasi and forms an isolated cartilaginous part. The homologization of these structures is not clear. The posteriorly situated part, connected to the tectum nasi, could be thought to be the cartilago cupularis. However, the narrow processes projecting far towards anterior can hardly be interpreted with certainty. They may be functionally conditioned, new secondary structures. De Burlet (1913a,b, 1914b), Schreiber (1916) and De Beer (1937) denote these structures in *Phocoena*, *Lagenorhynchus* and *Globicephala* simply as the tectum nasi. Eales (1950) denotes the structure situated in front of the tectum nasi in *Monodon* as the "isolated cartilage near [the] blow-hole". In our earlier paper (Klima and Van Bree 1985) on *Phocoena*, we have suggested that this may be the material for the future processus alaris superior. Apparently, these anlagen give rise to the variable cartilaginous structures that form a part of the closing system of the external nasal opening of adults. Although these cartilages have long been known to occur in adult Odontoceti (Von Baer 1826; Kükenthal 1893; Gruhl 1911), their exact description is still wanting and only recently attempts have been made at a more precise explanation of their function (Schenkkan 1973; Mead 1975; Cranford 1988; Amundin and Cranford 1990; Cranford et al. 1996).

The third type of the cupula nasi anterior is found in the sperm whale, *Physeter catodon*. As already mentioned (see Sect. 4.1.7), in *Physeter* the tectum nasi is transformed into a long and narrow bar that passes the whole soft anterior part of the head obliquely dorsally, accompanying the left nasal passage. This elongated tectum nasi is joined by the cupula nasi anterior, likewise in the form of a long and narrow, however paired, bar. The connection of the cupula nasi anterior sinistra to the tectum nasi has been preserved on the left-hand side. Later these structures fuse to form a unitary cartilaginous bar that remains preserved even in adults, to support the left nasal passage (see Sect. 7.4). Also, the ring of several cartilaginous parts, embracing the nasal aperture of the adult, probably develops from the material of the left cupula nasi anterior. The right-hand nasal passage, no longer functional for respiration in the sperm whale, is first accompanied by the cupula nasi anterior dextra, which, however, soon loses its connection to the tectum nasi and is evidently wholly reduced later (Behrmann and Klima 1985; Klima et al. 1986; Klima 1987a, 1990). The transformations of the cupula nasi anterior in *Physeter* are the most peculiar among those known to occur in all mammals. They clearly show how flexibly and variously the structures of the embryonic nasal capsule can change if required by a functional change.

6.2.6 Maxilloturbinale

In most mammals the position of the maxilloturbinale is relatively constant, and there in no problem in recognizing this structure as such. It is the lower edge of the paries nasi, which is inwardly bent or furled, and at the same time, it is the lowest boundary of the interior of the nasal capsule (Slabý 1987b). According to Gaupp (1908), the maxilloturbinale is a new structure in mammals. In monotremes it probably develops from an isolated anlage, as ascertained by Kuhn (1971) in *Tachyglossus;* whether this is so even in *Ornithorhynchus* is not quite clear (De Beer and Fell 1936). In all Marsupialia and Placentalia examined, the maxilloturbinale develops, from the very beginning, as a component of the paries nasi (e.g., *Perameles*: Cords 1915; *Erinaceus*: Fawcett 1918; *Rousettus*: Starck 1943; *Myotis*: Frick 1954; *Tupaia*: Spatz 1964; Zeller 1987; *Propithecus*: Starck 1966; *Pan*: Starck 1960; *Oryctolagus*: Voit 1909; Frick and Heckmann 1955; *Zaedyus*: Reinbach 1955, etc.). In Rodentia the maxilloturbinale is particularly extensive and its position may vary (cf. Fawcett 1917; Kadam 1972, 1973; Zimmermann 1976; Dierbach 1985; Frick 1986; Hauck 1987; Schrenk 1989). In Rodentia, Schrenk (1989) distinguishes three possible positions, depending on whether the origin of the maxilloturbinale from the paries nasi is situated more rostrally, in the

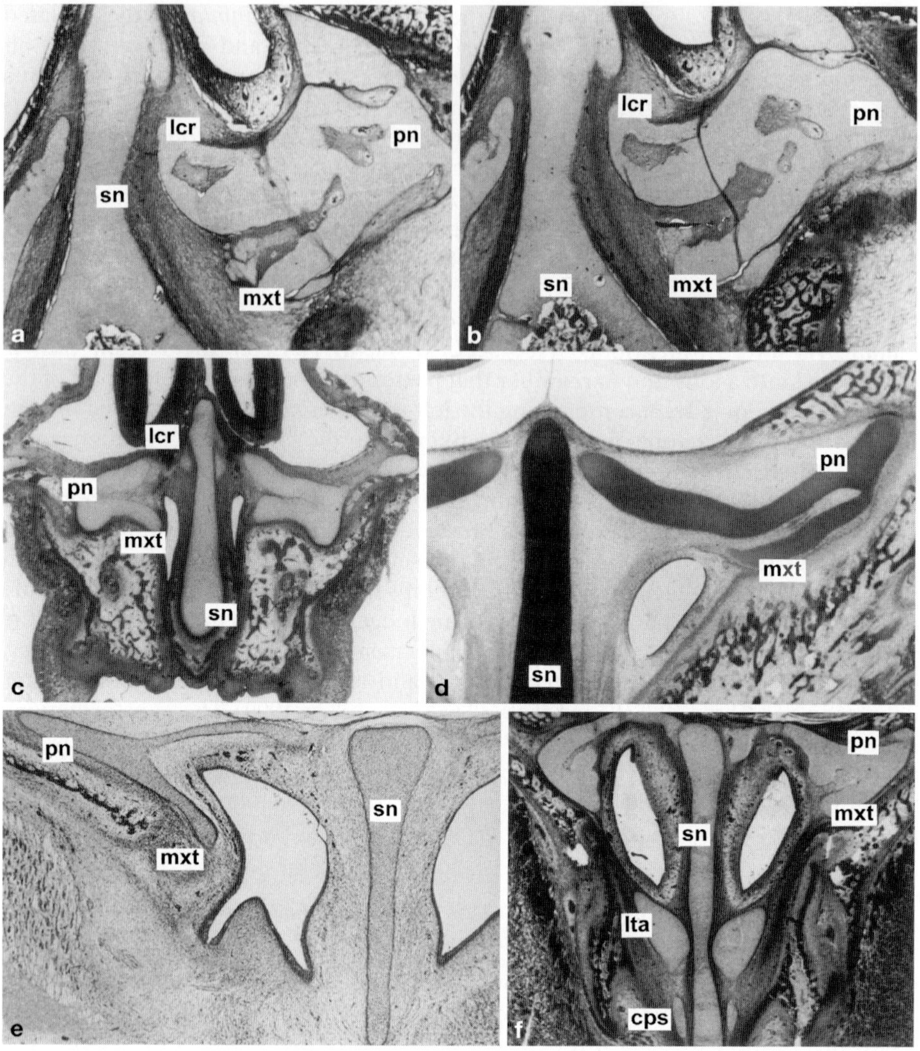

Fig. 43a-f. Maxilloturbinale (*mxt*) in horizontal sections. *Phocoena phocoena*, 167 mm TL, from above (**a**) to below (**b**). **c** *Physeter catodon*, 97 mm TL. **d** *Monodon monoceros*, 137 mm TL. **e** *Stenella attenuata*, 89 mm TL. **f** *Lagenorhynchus albirostris*, 107 mm TL

middle, or more caudally. Being a cartilaginous element of the nasal capsule, the maxilloturbinale is accompanied by the maxillare, a dermal element of the exocranium, all along its length.

An alleged maxilloturbinale in Cetacea was already reported at the beginning of the 19th century in connection with an attempt at homologizing the small bones lying at the anterior edge of the nasal aperture of adult *Phocoena* (Meckel 1825; Von Baer 1826; Mayer 1835). However, this purely hypothetical homologization was not based on any embryonic material. It was Kükenthal (1893) and Boenninghaus (1903) who were able to examine embryonic *Phocoena* for the purpose of clarifying the origin of these so-called Meckelian ossicles. However, they erroneously thought the maxilloturbinale was in the complex of the lamina transversalis anterior and the cartilago paraseptalis (cf. Sects. 6.3.1, 6.3.3, 7.5). The first correct description of a maxilloturbinale in Odontoceti was presented by Klima and Van Bree (1985) for *Phocoena*, *Lagenorhynchus* and *Monodon*. The first correct description of a maxilloturbinale in Mysticeti was presented by De Burlet (1914a) for *Balaenoptera*. In *Megaptera*, Honigmann (1917) found a thorough bar, which he denoted as the maxilloturbinale plus the crista semicircularis of the paries nasi.

In all embryonic cetaceans examined by us, the maxilloturbinale develops unequivocally as a part of the paries nasi. It forms an inward fold of its lower edge. In Mysticeti the maxilloturbinale is found in the form of a moderately arched furl that closes the inside of the nasal capsule (see Figs. 21, 25). The maxilloturbinale of *Physeter* is flattened, protruding from the paries nasi towards the interior and embracing the still well preserved nasal cavity (Fig. 43c). In the remaining Odontoceti there is hardly any cavity of the nasal capsule, and the maxilloturbinale does not protrude from the paries nasi, but is parallel to it, intimate with its anterior surface. It forms a fairly large, flattened plate (Fig. 43), the possible function of which is to strengthen, together with the paries nasi, the connection between the viscerocranium and the neurocranium of the Odontoceti.

6.2.7 The Nasal Cavity

The paired inner spaces of the nasal capsule form the left and right nasal cavity. This cavity reaches, on either side, from the cupula nasi anterior in front up to the cupula nasi posterior in the rear. Depending on the different functional requirements of different mammals, this space may vary extremely in shape and size. In general, this space is lined with olfactory mucosa, the area of which varies in dependence on the performance of the sense of smell. In order to increase this area, the inside relief is richly divided. On the one hand, variously formed projections protrude to the inside, the so-called turbinalia, either in the form of epithelial folds or as ledges provided with a skeletal strut. On the other hand, between these projections there are smaller or larger niches, the so-called recessus.

For a long time, attempts have been made at developing a general pattern of the nasal capsule that would be valid for all mammals and from which all the various modifications could be derived. The fundamental papers in this line have been presented by Nordenbos (1905), Gaupp (1908), Voit (1909) and Terry (1917). The paper by Reinbach (1952) has become authoritative and all subsequent authors who tackled this topic (Frick 1954; Kuhn 1971; Maier 1981; Schrenk 1989) have relied on it. According to Reinbach (1952), the inside of the nasal capsule can be divided into four parts, viz., (1) the pars anterior, (2) the pars lateralis, (3) the pars posterior, and (4) the paraseptal space. However, the boundaries between the individual parts, as well as the structures included in them, are not invariably unequivocally interpretable (cf. Schrenk 1989). This is true not only of the nasal capsule of various representatives of land mammals but particularly of the aquatic ones, in which the sense of smell is reduced, and consequently, even the nasal capsule between the land mammals and the

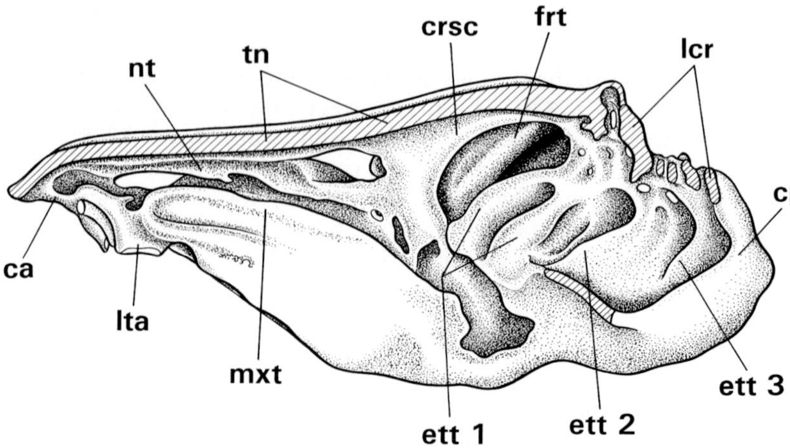

Fig. 44. Nasal cavity in a fetus of European rabbit, *Oryctolagus cuniculus*, viewed from the inside. (After Frick and Heckmann 1955)

aquatic cetaceans can only be based on a not very much modified nasal skull, such as that of *Oryctolagus* (Fig. 44).

In the baleen whales examined, *Balaenoptera* (Fig. 45a,b) and *Megaptera* (Fig. 45c,d), the entrance to the inside is formed by the cupula nasi anterior and the anterior vault of the paries nasi, the exit by the cupula nasi posterior. The upper edge of the nasal cavity consists of the tectum nasi in front and of the lamina cribrosa in the rear. The lower edge is closed by the maxilloturbinale. The inner relief of the nasal cavity shows the following structures in the antero-posterior direction: the crista semicircularis, the recessus frontoturbinalis, the frontoturbinale, the recessus maxillaris, the ethmoturbinale I and the recessus ethmoturbinalis. As regards the formation of the inner relief, there are only inappreciable differences between the two genera examined. On the whole, there is a heavy reduction of the relief structures, especially in the posterior section of the nasal cavity, with only the ethmoturbinale I being present as a skeletal structure of all other turbinals. The remaining smaller folds in this space are only formed by the mucosa, and they are irregular to such an extent that they can hardly be homologized.

Our observations agree largely with those on *Balaenoptera* by De Burlet (1914a) and *Megaptera* by Honigmann (1917). However, they differ considerably in the interpretation of the findings and particularly in the evaluation of the results. De Burlet (1914a) has attempted to interpret the posterior epithelial folds in *Balaenoptera* as the ethmoturbinale II and ethmoturbinale III. At the same time, his paper contains no firm support for this interpretation, either in the photographs of the successive histological sections or in the spatial reconstruction illustrated by him. His opinion has apparently led Honigmann (1917) to a similar interpretation of the posterior epithelial folds in *Megaptera* as the ethmoturbinale II and ethmoturbinale III. At the same time, Honigmann (1917) has partly mistaken the ethmoturbinale I for the maxilloturbinale. Based on this not quite correct interpretation by De Burlet (1914a) and Honigmann (1917), De Beer (1937) has arrived at a generalizing statement that "...internally, the capsule contains only those turbinals typical of the hinder region." This must

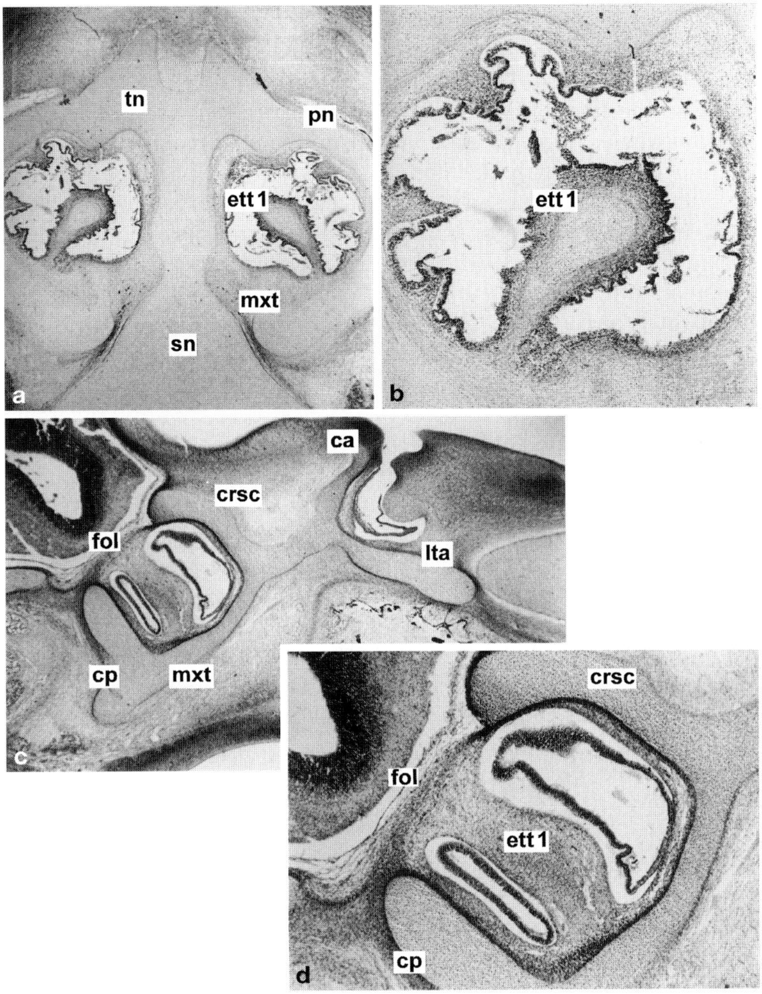

Fig. 45a-d. Nasal cavity. *Balaenoptera physalus*, 156 mm TL, transverse section in overall view (**a**) and in detail (**b**). *Megaptera novaeangliae*, 107 mm TL, sagittal section in overall view (**c**) and in detail (**d**)

necessarily lead to the idea that the nasal cavity of baleen whales is modified only in its anterior part, but remains unchanged posteriorly, corresponding to the relations found in land mammals. In contrast to that, the nasal cavity of toothed whales should be much more and differently reduced.

The overemphasized similarity between baleen whales and land mammals on the one hand, and the differences between the baleen and the toothed whales on the other had already led Kükenthal (1893) to the opinion that toothed and baleen whales are not directly related and that the similarity in the structure of their organs are mere convergences. On the basis of a comparison of the embryonic nasal cavity, Honigmann (1917) arrived at a still more far-fetched opinion stating that two different animal phyla ("Tierstämme") are involved. Thus, it is the very nasal cavity that has repeatedly been employed as one of the keystones of that systematic classification according to which the cetaceans do not form a natural unit, being of diphyletic origin (Slijper 1962; Yablokov 1964; Kuzmin 1976; Rice 1977, 1984) (cf. Chap. 9).

However, comparing the inside spaces of the nasal capsule in greater detail, it appears that the baleen whales differ considerably from the land mammals and that they show many traits in common with the toothed whales. Of the wide range of mostly large turbinalia of land mammals, only a few have remained preserved in the baleen as well as toothed whales, and these only as small rudimentary structures. In the baleen whales, this is only the maxilloturbinale with the associated crista semicircularis, the indistinct frontoturbinale and, finally, the ethmoturbinale I. In the toothed whales, this is only the maxilloturbinale and a hardly distinguishable ethmoturbinale I. Thus, it is a case of reduction or preservation of the same structures in the baleen and the toothed whales and, hence, a case of an identical reconstruction mode of the original nose of the land mammals. Such an accordance can in no way be considered to be a convergent adaptation in two mutually unrelated animal groups.

However, the nasal cavity is more heavily reduced in the toothed whales than in the baleen ones. This pertains not only to the above-mentioned turbinalia but even the whole space is more heavily narrowed. A true space is only found in *Physeter* (cf. Klima et al. 1986), whereas in all other Odontoceti only a narrow fissure remains. Also, the whole nasal cavity of the Odontoceti is much more steeply oriented than in the Mysticeti. Besides, there are differences in the condition of the olfactory mucosa as well as in the form of the lamina cribrosa (cf. Sect. 6.2.2). The structure described as the nasoturbinale in *Phocoena* by Kükenthal (1893) is in fact the lamina transversalis anterior (cf. Sect. 6.3.1), and it has no relation whatever to the inside space of the nasal capsule.

6.2.8 Processus paraseptalis posterior

In most mammals the posterior part of the paries nasi juts out as an arched cupula nasi posterior. It forms the caudal closure of the nasal cavity and, at the same time, a narrowing for the nasal passage. Dorsally, it is connected with the tectum nasi. Ventrally, it is mostly inclined towards the septum in the form of an incrassate plate, the lamina transversalis posterior, without being connected to the septum, however. From the anterior edge of this plate, a spine-like processus paraseptalis posterior may project anteriorly. This projection may correspond to the posterior part of the originally unitary cartilago paraseptalis communis, the anterior part of which has separated as the proper cartilago paraseptalis (cf. Sect. 6.3.3). Like the cartilago paraseptalis, even the processus paraseptalis posterior is accompanied from below by the dermal bone, the vomer, all along its length. This generally valid pattern, however, may vary in different mammalian groups, as pointed out by several authors (cf. De Beer 1937; Reinbach 1952; Frick 1954; Slabý 1960, 1987b; Kuhn 1971; Starck 1979; Maier 1987; Schliemann 1989, 1991; Schrenk 1989).

In Cetacea, the cupula nasi posterior forms a thickened margin at the posterior edge of the paries nasi. In its lower, slightly inwardly bent part, one could recognize, in Mysticeti, the lamina transversalis posterior (De Burlet 1914a). In Odontoceti this floor plate is less distinct and hardly definable from the proper cupula nasi posterior. The

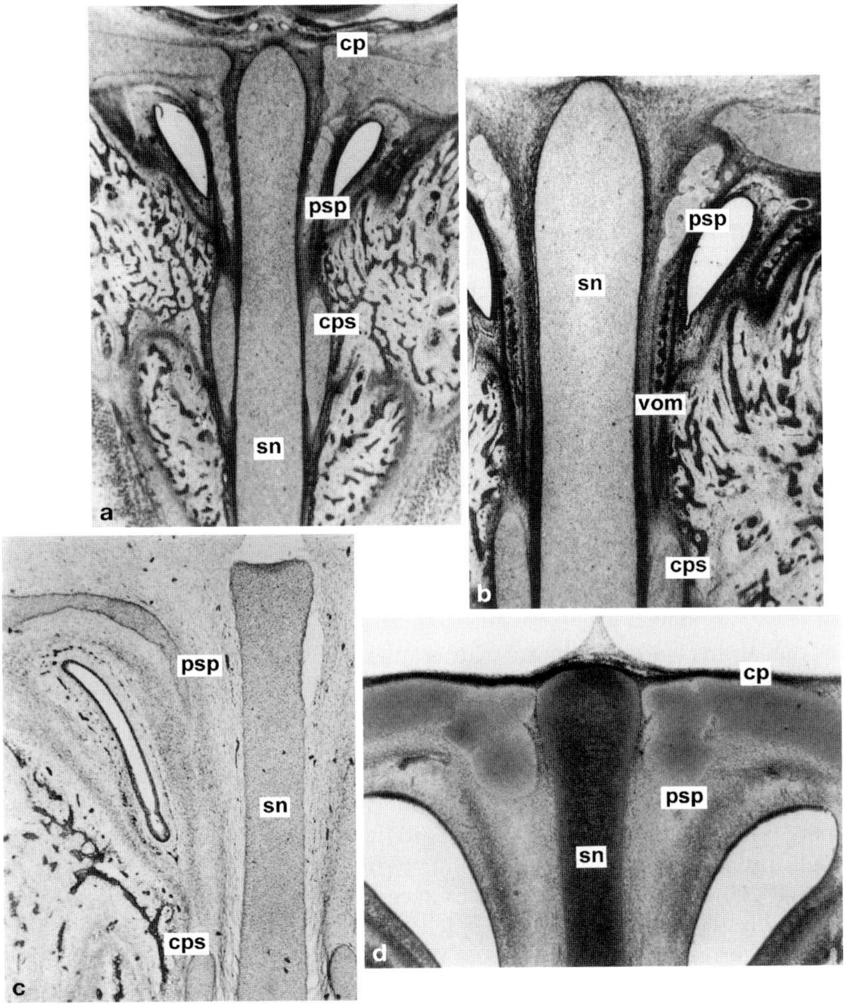

Fig. 46a-d. Processus paraseptalis posterior (*psp*) in horizontal sections. *Lagenorhynchus albirostris*, 107 mm TL, from above (**a**) to below (**b**). **c** *Stenella attenuata*, 89 mm TL. **d** *Monodon monoceros*, 137 mm TL

processus paraseptalis posterior (Fig. 46) is the only distinctly identifiable structure in this region in all cetaceans examined. The processus rises like a recurrent projection from the lower medial edge of the cupula nasi posterior. It becomes intimate with the septum nasi, without fusing with it. Its lower edge is supported by the vomer.

The first denomination of a processus paraseptalis posterior in a cetacean embryo has been provided by De Burlet (1913a) for a *Phocoena*. Later, De Burlet (1914b) found this projection in a *Lagenorhynchus*, in which case it is particularly large, reaching far anteriorly and, at the same time, consisting only of perforated porous cartilage. This unusual development of the processus paraseptalis posterior probably led De Burlet (1914b) to the erroneous opinion that this was the only cartilago paraseptalis in

Lagenorhynchus and that, in this case, the anterior structures included no cartilago paraseptalis. De Beer (1937) corrected this error and labeled, in the illustration adopted from De Burlet (1914b), this projection as the posterior paraseptal cartilage. A fairly big processus paraseptalis posterior has also been found in *Globicephala* (Schreiber 1916; De Beer 1937).

In our material of the two representatives of Mysticeti, *Balaenoptera* and *Megaptera*, we have found only a feebly developed processus paraseptalis posterior. Among the Odontoceti, this projection is particularly poorly developed in *Physeter* (Klima et al. 1986), and in *Monodon* and *Phocoena* (Klima and Van Bree 1985) there is only a relatively small projection. On the contrary, *Stenella* and *Delphinus* possess a well-developed processus paraseptalis posterior, and the structure is the most developed in *Lagenorhynchus* (Fig. 46a,b).

6.3 The Anterior Side Wall Structures

6.3.1 Lamina transversalis anterior

Originally, the lamina transversalis anterior is an exclusive element of the nasal floor, as can still be observed in many reptiles (Gaupp 1900; De Beer 1937; Slabý 1986, 1987b). In terrestrial mammals it is laterally enlarged and upturned, so that it consists medially of a horizontal and laterally of a vertical section (Sturm 1936; De Beer 1937). The horizontal section also participates in the formation of the nasal floor, in which case it can connect with the septum nasi. The vertical section can topographically no longer belong to the floor elements, as it participates in the formation of the side wall of the nasal capsule and is mostly connected with the paries nasi.

A thorough connection of the lamina transversalis anterior with the paries nasi dorsally and with the septum nasi ventrally leads to a ring-like structure surrounding the nasal cavity, termed the “zona anularis” by Gaupp (1908). It is uncertain whether a complete zona anularis is primitive and a divided one is advanced. The conditions may vary considerably without any appreciable evolutional and functional connections, and thus also the concept of the zona anularis has been interpreted very differently (De Beer 1937; Starck 1941; Frick 1954; Spatz 1964). It is a purely formal denomination. Kuhn (1971) suggests that the term “zona anularis” should be abandoned. On the contrary, Schliemann (1987) thinks that “...since the anterior part of the nasal cavity is enclosed around its entire circumference by cartilage in many Theria, at least in certain stages, the term zonula anularis continues to be applicable.”

In Monotremata (*Tachyglossus*: Gaupp 1908; Kuhn 1971; *Ornithorhynchus*: De Beer and Fell 1936) the lamina transversalis anterior forms an extensive, massive cartilaginous plate. Even in Marsupialia (*Perameles*: Cords 1915; Esdaile 1916; *Dasyurus*: Fawcett 1917: *Didelphis*: Toeplitz 1920) the lamina transversalis anterior is said to be generally much stronger than in Placentalia, which, according to Kuhn (1971) may be connected with the special mechanical requirements of the snout tip during an earlier developmental period. A similar strengthening and temporary oversized development of certain structures of the embryonic cartilaginous skeleton has been observed, in connection with the extreme conditions of the premature birth in Marsupialia, even in their shoulder girdle (Klima 1987b; Klima and Bangma 1987).

In Placentalia the lamina transversalis anterior is usually smaller and above all, much narrower. Particularly in some Rodentia (*Bathyergus*, *Georychus*, *Ctenodactylus*: Schrenk 1989) can it be present only in the form of a thread-like cord. The reduction of the lamina transversalis anterior can affect the horizontal or the vertical section to different extents. A massively developed horizontal part with a heavily reduced vertical one could, according to Schliemann (1987), be considered to be a plesiomorph character in mammals. Such a situation is found, e.g., in some Insectivora (*Talpa*: Fischer 1901; *Erinaceus*: Fawcett 1918a), Xenarthra (*Dasypus*, *Zaedyus*: Reinbach 1952, 1955) and Carnivora-Fissipedia (*Canis*: Olmstead 1911; Schliemann 1966, 1987, 1991; *Genetta*, *Proteles*: Frangioni 1970; *Cryptoprocta*, *Paradoxurus*: Köhncke 1980, 1985). A quite contrary situation, with a strongly developed vertical and considerably reduced horizontal part, is found, e.g., in some Insectivora (*Solenodon*: Menzel 1979), Chiroptera (*Rousettus*: Starck 1943; *Myotis*: Frick 1954), Carnivora-Fissipedia (*Felis*: Terry 1917; Timm 1987; Timm et al. 1989; *Herpestes*: Sturm 1985), Lagomorpha (*Oryctolagus*: Voit 1909; De Beer and Woodger 1930; Frick and Heckmann 1955), Artiodactyla (*Bos*: Sturm 1937; *Camelus*: Slabý 1987a; *Cervus*, *Dama*: Slabý 1990a,b), etc. In Carnivora-Pinnipedia (*Leptonychotes*: Fawcett 1918b; *Mirounga*: Kummer and Neiss 1957) the lamina transversalis anterior is almost completely reduced, being "...represented only by processes of the paries and widenings of the septum nasi" (Schliemann 1987). In Scandentia (*Tupaia*: Henckel 1927; Spatz 1964; Zeller 1987) as well as in most primates (Wen 1930; Schultz 1935; Starck 1975a,b; Maier 1981) the lamina transversalis anterior is present, but in Hominoidea (*Pan*: Starck 1960; *Homo*: Fawcett 1917; Reinbach 1963a; Bersch and Reinbach 1970) it is completely reduced, probably due to the fact that their nasal capsule is fundamentally altered by being moved from the originally praecerebral to a subcerebral position (Starck 1975b).

In ontogeny the lamina transversalis anterior mostly grows out of the paries nasi, e.g., in some Marsupialia (*Didelphis*: Levi 1909), Chiroptera (*Rousettus*: Starck 1943), Scandentia (*Tupaia*: Zeller 1987), Rodentia (*Otomys*, *Rhabdomys*: Eloff 1948, 1951a; *Ctenodactylus*: Schrenk 1989; *Sciurus*: Slabý 1991), Lagomorpha (*Oryctolagus*: De Beer and Woodgeer 1930). In *Ornithorhynchus* (Monotremata, De Beer and Fell 1936) and in *Tatera* (Rodentia, Kadam 1972/73) the lamina transversalis grows out in a contrary direction, i.e., from the septum nasi towards the side wall. In *Tachyglossus* (Monotremata, Kuhn 1971) the lamina transversalis anterior develops from isolated paired cartilaginous centers. According to Kuhn (1971), ontogeny provides no clue as to the homology of this structure.

The connections of the lamina transversalis anterior with other elements of the embryonic nasal skull may be diverse (cf. Sturm 1936; Frick 1954; Schliemann 1987; Schrenk 1989). Connections with the cartilago paraseptalis and the processus lateralis ventralis are regularly observed, as is often also a connection with the cartilago ductus nasopalatini and the cartilago palatini. Occasionally, there occur contacts with other cartilaginous structures. From among the elements of the dermatocranium, a close connection with the lamina transversalis anterior is shown by the praemaxillare, which becomes intimate with the lateral surface of the cartilage and accompanies it all along its length.

In his examinations of cetaceans, Kükenthal (1893) recognizes a lamina transversalis anterior neither in *Balaenoptera* nor in *Phocoena* and an "embryonic dolphin" although it is distinctly seen in the illustrated cross sections of all three species. The author considers the anterior part of this structure to be the "lateral nasal wall", the posterior one, to be partly the "nasoturbinale" and partly the "maxilloturbinale". Boenninghaus (1903) simply denotes the complex of the cartilago paraseptalis and the lamina transversalis anterior of *Phocoena* 48 mm, and even in the larger embryonic *Phocoena* 92 mm, he is not quite sure what structures are present in this region. He writes that the paraseptal cartilage is seen in connection with a cartilaginous part lying dorsal of it, the interpretation of which is difficult. It is questionable if this may be an original section of the floor of the nasal capsule, corresponding to the lamina transversalis anterior, or if it should be considered a remainder of the lateral nasal wall. In his subsequent papers on *Balaenoptera* and *Lagenorhynchus*, De Burlet (1914a,b) already recognized the lamina transversalis anterior and described it as such. Also, Honigmann (1917) supports the opinion of De Burlet and believes to have recognized a lamina transversalis anterior in *Megaptera;* however, he also denotes the paraseptal cartilage to be the lamina transversalis anterior. De Beer (1937) adopted the results of the two authors as well as those obtained from *Globicephala* by Schreiber (1916) and he attempted to correct their different interpretations. Nevertheless, even his own interpretation is not exact. Thus, the connections between the lamina transversalis anterior and other cartilaginous structures, such as the paries nasi, the cartilago paraseptalis and the cartilago ductus nasopalatini, remain rather uncertain in the comparison by De Beer (1937). The processus lateralis ventralis is not mentioned at all and even the characteristic relationship of the lamina transversalis anterior to the praemaxillare of the dermatocranium remains unnoticed. Additional confusion is due to the inconsistency between the lamina transversalis anterior illustrated and indicated in *Globicephala* and the statement in the text that "...there is no lamina transversals anterior." The inexact indicating lines in the illustrations by De Beer (1937), marking only the protrusion as the lamina transversalis anterior, which, in fact, corresponds to the processus lateralis ventralis, has apparently led even Eales (1950) to the same erroneous interpretation of the lamina transversalis anterior in her reconstruction of the chondrocranium of *Monodon.*

The lamina transversalis anterior is present in all cetaceans examined so far (Klima and Van Bree 1985; Klima et al. 1986; Klima 1987a; the present paper). It has the shape of a slender bar bent to form a quarter-ring, lying mostly horizontal and surrounding the vertical nasal passage from the lateral and rostral sides. Together with the paries nasi, the tectum and the septum, it forms a ring which is not quite closed, however, as the lamina transversalis anterior, while being intimate with the tectum nasi anteriorly, does not fuse with it in any of the species or at any developmental stage. Consequently, there is no zona anularis. Posteriorly, the lamina transversalis anterior connects continuously with the paries nasi. Exceptions are found in *Phocoena*, with a gap between the two structures on either side in all developmental stages, and *Physeter*, in which, in connection with the marked asymmetry of the skull, the two structures are mutually connected on the right side but remain separated on the left one. In ontogeny the earliest anlage of the lamina transversalis anterior grows out of the paries nasi. Exceptions even in this point are found in *Phocoena* (bilaterally) and *Physeter* (left side), in which the lamina transversalis anterior develops from independent mesenchymatous centers. The anterior section of the lamina transversalis anterior, situated directly next to the septum, is more robust than the posterior one, which mostly consists of only a narrow connecting strip to the paries nasi. The rostral end of the anterior section is connected, in some genera (*Stenella*, *Delphinus*, *Lagenorhynchus*, *Monodon*),with the processus lateralis ventralis; below, it continues in the cartilago paraseptalis in all genera. Connections with other cartilaginous elements have not been observed. Of the desmal elements, the praemaxillare becomes intimate with the lamina transversalis anterior from the rostral and lateral sides and accompanies it all along its length (Figs. 47–48).

6.3.2 Processus lateralis ventralis

From the lower edge of the septum nasi in a lateral direction, a short connecting shaft reaches to the anterior edge of the nasal floor. As far as it has been recognized at all as a distinct structure, it has been denoted as the processus lateralis ventralis. According to Sturm (1936), it is in fact no independent element but a ventral projection of the cupula nasi anterior. Depending on the development of the anterior nasal cupula, however, this original connection may disappear, and the processus lateralis ventralis joins the lamina transversalis anterior. This is also the case in most land mammals, e.g., Chiroptera (*Myotis*: Frick 1954), Scandentia (*Tupaia*: Zeller 1987), Primates (*Homo*: Reinbach 1963; Bersch and Reinbach 1970; *Saimiri*, *Pithecia*: Maier 1981, 1991; *Saguinus*: Müller 1981; *Galago*: Warich 1986; *Daubentonia*, *Homo*: Maier 1991), Xenarthra (*Dasypus*: Reinbach 1952), Rodentia (*Cavia*: Dierbach 1985; *Bathyergus*, *Georychus*: Schrenk 1989; *Sciurus*: Slabý 1991), etc. In Artiodactyla (*Bos*: Sturm 1936; *Cervus*, *Dama*: Slabý 1990a,b), the processus lateralis ventralis can apparently reach up over the cartilago paraseptalis. The processus lateralis ventralis is not present in many Placentalia examined, and it is also absent in Monotremata (*Ornithorhynchus*: De Beer and Fell 1936; *Tachyglossus*: Kuhn 1971). However, the processus lateralis ventralis of Theria (Marsupialia, Placentalia) may be homologous with the crista marginalis, a derivative of the cupula nasi anterior of Prototheria (Monotremata), as assumed by Kuhn (1971). Among other things, this assumption would be supported by the close relations of the two cartilaginous structures to the desmal praemaxillare.

In Cetacea the processus lateralis ventralis was first found in *Lagenorhynchus* by De Burlet (1914b). His illustration shows this structure as a small, isolated cartilaginous part close to the anterior edge of the lamina transversalis anterior. However, De Burlet (1914b) was not quite sure with which element of the nasal capsule this structure may be homologous, and he mentions the possibility that it may be the cartilago ductus nasopalatini. De Beer (1937), adopting De Burlet's (1914b) reconstruction of *Lagenorhynchus* for his monograph, did not label this structure in the illustration and did not mention it at all in the text. Only after this structure had been examined in another embryonic *Lagenorhynchus* as well as in *Monodon*, and after the metamorphosis of the nasal structures in *Phocoena* had been clarified, was this cartilaginous part interpreted as the processus lateralis ventralis with considerable certainty (Klima and Van Bree 1985; Klima 1987a). This concept has also been confirmed by the findings in *Stenella* and *Delphinus*.

In Cetacea the processus lateralis ventralis is situated exactly the same as in land mammals, i.e., at the anterior edge of the lamina transversalis anterior, and it can partly reach up over the cartilago paraseptalis (Figs. 47, 48). It has no connection with the

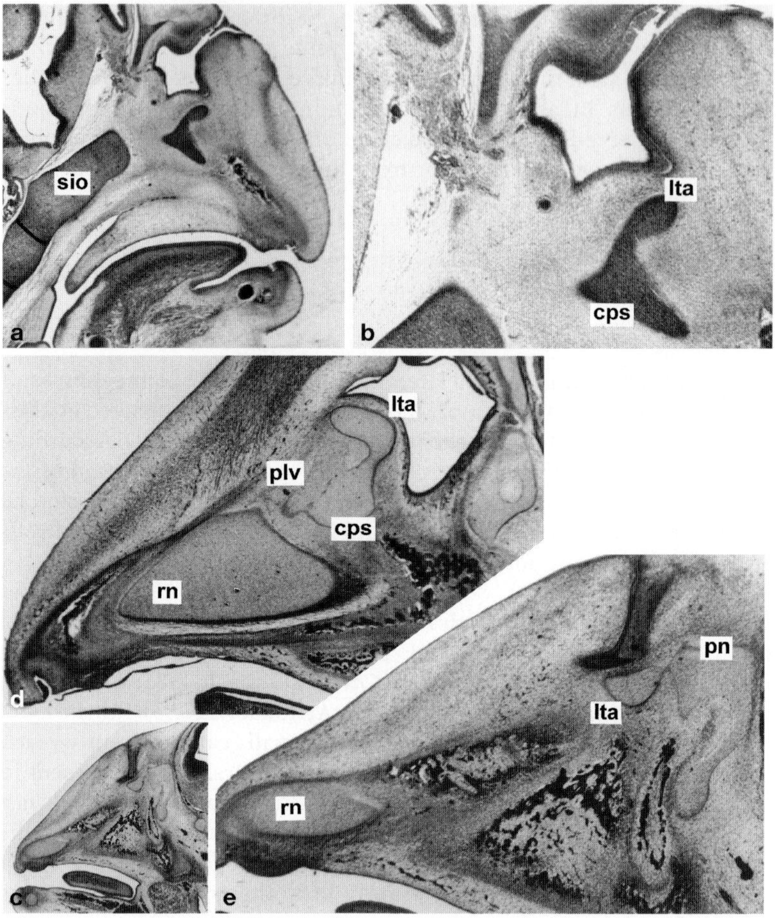

Fig. 47a-e. The anterior side wall structures of the nasal skull, lamina transversalis anterior (*lta*), processus lateralis ventralis (*plv*) and cartilago paraseptalis (*cps*) in sagittal sections. *Phocoena phocoena*, 60 mm TL, in overall view (**a**) and in detail (**b**). *Phocoena phocoena*, 95 mm TL, in overall view (**c**) and in detail from medial (**d**) to lateral (**e**)

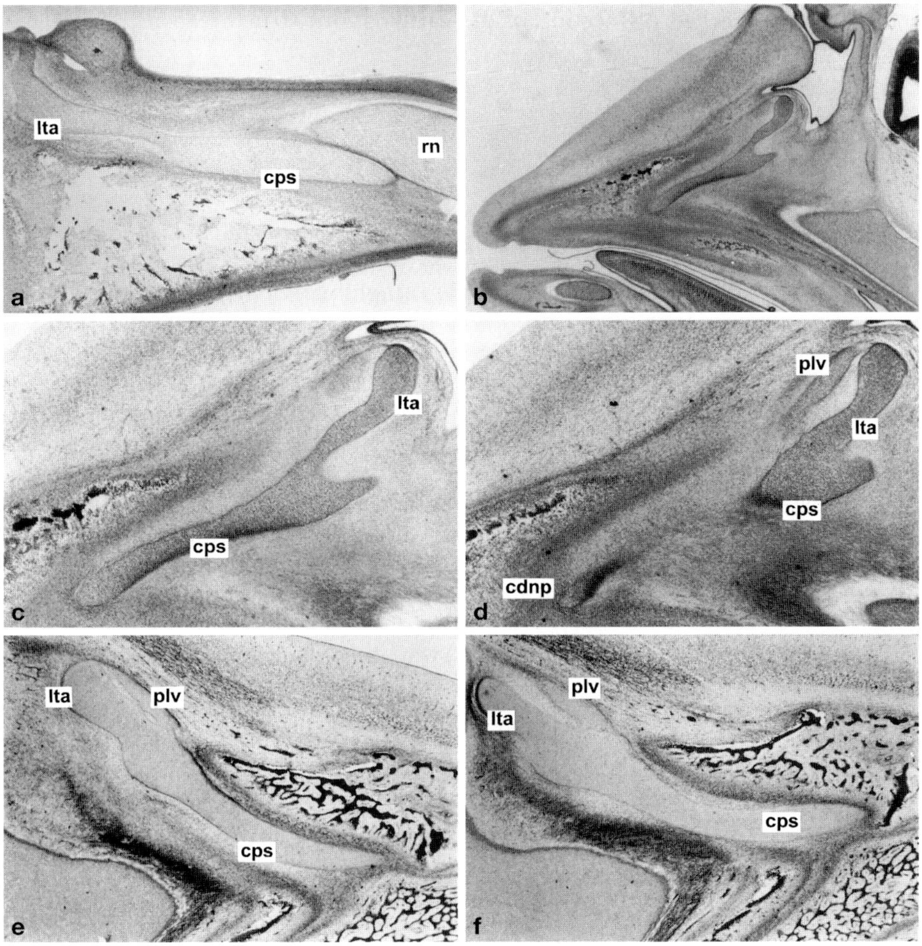

Fig. 48a-f. The anterior side wall structures of the nasal skull, lamina transversalis anterior (*lta*), processus lateralis ventralis (*plv*), cartilago paraseptalis (*cps*) and cartilago ductus nasopalatini (*cdnp*) in sagittal sections. **a** *Megaptera novaeangliae*, 107 mm TL. **b** *Stenella attenuata*, 85 mm TL, in overall view, and in detail from medial (**c**) to lateral (**d**). **e** *Delphinus delphis*, 113.5 mm TL, from medial, to lateral (**f**)

septum nasi any longer. At early developmental stages, it forms a continuum with the lamina transversalis anterior but can separate from it partly or wholly at later stages (*Stenella*, *Delphinus*). This may also be true of *Lagenorhynchus* and *Monodon*, genera of which no early stages have been available and in which, thus, the processus lateralis ventralis has only been observed as a partly isolated element. In some genera (*Globicephala*, *Phocoena*) it is only visible as a distinctly protruding tubercle at the anterior edge of the lamina transversalis anterior. In others (*Physeter*, *Balaenoptera*, *Megaptera*) it has probably been completely integrated, without any visible delimitation, in the anlage of the lamina transversalis. In all genera examined there is, in this place, a marked connection of the cartilaginous structures with the dermal praemaxillare.

6.3.3 Cartilago paraseptalis

In land mammals, the occurrence of the cartilago paraseptalis can assume very different forms. The cartilage develops as a derivative of the medial wall of the original paired nasal capsule, as can still be observed in amphibians and reptiles (Gaupp 1893, 1900; De Beer 1937; Slabý 1960, 1986, 1991). In mammals this original connection is usually lost. There develops an independent cartilago paraseptalis communis which accompanies the septum nasi from lateral and reaches anteriorly from the lamina transversalis anterior up to the lamina transversalis posterior or to the cupula nasi posterior (Slabý 1960; Kuhn 1971). Functionally, the posterior part of the cartilago paraseptalis supports the primary palatal region, its anterior part embracing Jacobson's organ (Starck 1967). It is questionable whether a completely preserved cartilago paraseptalis communis, passing all along the nasal floor, actually corresponds to the original primitive condition, as thought by most authors. Very diverse opinions on this point are found particularly in the earlier literature (Seydel 1899; Gaupp 1908; Zuckerkandel 1909; Fawcett 1911, 1918; Toeplitz 1920; Sturm 1936). A detailed discussion of this problem is found in Kuhn (1971) who illustrates several possible transformations of the cartilago paraseptalis and arrives at the following conclusion: In recent representatives, one can find a diversity of reductions of this cartilage, which cannot be expressed in a strict scheme and which have been reached in many parallel lines.

A thorough cartilago paraseptalis communis is found, for example, in some Marsupialia (*Trichosurus*: Broom 1909; *Dasyurus*: Esdaile 1916; *Monodelphis*: Maier 1987), Primates (*Alouatta*: Starck 1941), Xenarthra (*Bradypus*: Schneider 1955), Rodentia (*Xerus*: Fawcett 1923; *Castor*: Karger 1988; *Sciurus*: Slabý 1991), Lagomorpha (*Oryctolagus*: Voit 1909), Sirenia (*Halicore*: Matthes 1921; *Trichechus*: Hirschfelder 1937), etc.

More often the cartilago paraseptalis consists of two parts. The anterior one is mostly connected with the lamina transversalis anterior, the posterior one with the lamina transversalis posterior or with the cupula nasi posterior. In case the two parts are present as relative large and distinctly delimited structures, they are denoted as the cartilago paraseptalis anterior and the cartilago paraseptalis posterior. In case the posterior part is reduced to only a small projection at the lower edge of the cupula nasi posterior, it is named the processus paraseptalis posterior, and the larger anterior part is denoted the cartilago paraseptalis anterior or simply the cartilago paraseptalis proper.

Such an anterior cartilago paraseptalis is found, for example, in Monotremata (Kuhn 1971), most Insectivora (Roux 1947), some Chiroptera (*Miniopterus*: Fawcett 1919; *Rousettus*: Starck 1943; *Myotis*: Frick 1954), Scandentia (*Tupaia*: Spatz 1964; Maier 1981; Zeller 1987), Primates (*Homo*: Macklin 1914, 1921; Slabý 1960; Reinbach 1963; *Pan*: Starck 1960; *Cercopithecus*: Starck 1967; *Galago*, *Saimiri*, *Pithecia*: Maier 1981; *Saguinus*: Müller 1981) Xenarthra (*Dasypus*: Reinbach 1952), Rodentia (*Erethizon*: Struthers 1927; *Otomys*: Eloff 1948; *Rhabdomys*: Eloff 1951a; *Rattus*: Youssef 1966; *Tatera*: Kadam 1972, 1973; *Cavia*: Rajtova 1972a; Dierbach 1985;

Mesocricetus: Rajtova 1972b; *Mus*: Frick 1986; *Ctenodactylus*: Schrenk 1989), Carnivora (*Leptonychotes*: Fawcett 1918; *Canis*: Schliemann 1966, 1987), and Artiodactyla (*Camelus*, *Dama*, *Cervus*, *Ovis*: Salbý 1987a, 1990a,b).

The two forms of the cartilago paraseptalis mentioned above, one as the complete "communis" and the other as the divided "anterior and posterior", are in fact not basically different forms. Also, they can be considered to be species-specific characters only with some limitations as, under certain conditions, they may be subject to heavy ontogenetic transformations. Thus, Schliemann (1987) found, in the early stages of some Carnivora (*Herpestes*, *Felis*), a united cartilago paraseptalis communis, and a twofold one in the later stages. Schliemann (1987) writes this: "The presence of a cartilago paraseptalis communis, existing only for a short period of ontogeny in *Felis*, clearly demonstrates that the concept of the 'stadium optimum' in the ontogeny of the mammalian skull is not valid. As in *Felis* a cartilago paraseptalis communis might also be found in other Carnivora on the basis of intermediate stages." That all parts of the chondrocranium are never fully developed at one and the same time and that, therefore, the idea of a stadium optimum in the development of the chondrocranium is fiction has already been pointed out by Starck (1975a) in another connection. Individual differences may also play a part in the development of the cartilago paraseptalis, as ascertained by Köhncke (1985) in *Cryptoprocta* (Carnivora). Zeller (1987) investigated the metamorphosis of the cartilago paraseptalis during ontogeny in *Tupaia* (Scandentia) and found considerable differences in its relative size, shape and relationships to other structures. He writes: "Only by looking at craniogenesis as a continuous morphogenetic process is it possible to distinguish between a species-specific character and a character which is characteristic of only a transitory stage of development or only of an individual embryo. This distinction is necessary for the comparative analysis of the mammalian skull. "

Although the anterior, rostral part of the paraseptal cartilage has found in contrast to the caudal part, an important new function in the earliest mammals in supporting the Jacobson organ (Kuhn 1971) and thus remains preserved much more often than the posterior part, its occurrence is not necessarily bound to the presence of the Jacobson organ. A massive cartilago paraseptalis anterior can be present even in mammals which posses no Jacobson organ at all, e.g., in some Chiroptera (*Rousettus*: Starck 1943; *Myotis*: Frick 1954), Carnivora (*Leptonychotes*: Fawcett 1918), etc. This is so in all Cetacea as well.

Wherever there is a close connection between the Jacobson organ and the paraseptal cartilage, the latter forms an open flute which, at later stages, can even close to form a tube, as e.g., in some Rodentia (*Cavia*: Rajtova 1972a; *Mus*: Frick 1986; *Ctenodactylus*: Schrenk 1989). In Monotremata (*Tachyglossus*, *Ornithorhynchus*: Kuhn 1971) this tube is additionally equipped with a cartilaginous turbinale. In many Marsupialia (Broom 1896; Toeplitz 1920), Insectivora (Roux 1947), Primates (Maier 1981) as well as some other mammals there is, laterally at the upper edge of the paraseptal cartilage, behind the place where it connects with the lamina transversalis, a small cartilaginous bar, denoted as the "outer bar" (Broom 1896) or the "fibula reuniens" (Reinbach 1952). According to Broom (1896), the presence or absence of this structure may be an important evolutional character in craniogenesis in mammals, a hypothesis which has been doubted by most subsequent authors (cf. discussions in Kuhn 1971; Schliemann 1987; Zeller 1987).

Originally a part of the primary chondral palate, the cartilago paraseptalis is replaced in mammals by the vomer, an element of the secondary desmal palate. During craniogenesis, the paraseptal cartilage is the first to appear but at the late stages it is either replaced by, or integrated into, the vomer, which can even take place within the histological process of ossification, as noted by Kuhn (1971). These histological processes have been particularly objectively illustrated by Starck (1967, 1975) as an overlapping of the vomer on the paraseptal cartilage. The close connection of these two structures has been reported several times (Zuckerkandl 1908; Fuchs 1909; Toeplitz 1920; Fawcett 1918; Sturm 1936; Eloff 1952; Reinbach 1955).

The cartilago paraseptalis is distinctly recognized in the cross sections through the nasal skull of embryonic *Balaenoptera*, *Phocoena* and an unidentified embryonic dolphin illustrated by Kükenthal (1893). However, Kükenthal himself considers this structure to be a part of the septum nasi. He thinks that the cartilago paraseptalis, i.e., the cartilago Jacobsoni, should be looked for somewhere in the vicinity of the outer nasal aperture, and he denotes as such a part of the cupula nasi anterior in *Balaenoptera.* According to Kükenthal, the cartilago paraseptalis may not occur at all in toothed whales. In *Phocoena*, Boenninghaus (1903) denotes the complex of anterior structures, in which also the cartilago paraseptalis is found, simply as the cartilaginous side plate. This lateral plate has also been described in *Phocoena* by De Burlet (1913a). While he avoids any labeling in his illustration, he writes in the text the following about this structure: In the flattened cartilaginous part lying along the septum we can apparently see the cartilago paraseptalis. De Beer (1937), who has adopted De Burlet's illustration for his monograph, denotes this cartilaginous structure as the paries nasi, which is certainly wrong. In a larger embryonic *Phocoena*, De Burlet (1913b) found a distinctly twofold side plate, the lower part of which he correctly identified with the paraseptal cartilage. According to De Burlet (1914b), the cartilago paraseptalis is absent from the complex of the anterior side wall structures in *Lagenorhynchus*, as it has shifted posteriorly and forms the massive processus paraseptalis posterior as a projection of the posterior margin of the nasal capsule. Nevertheless, De Beer (1937) has described, in the reconstruction of the chondrocranium of *Lagenorhynchus* adopted from De Burlet (1914b), a cartilago paraseptalis in a part of the anterior side wall structures. According to De Beer (1937), the long cartilaginous bar in the side wall structures of *Lagenorhynchus*, *Phocoena*, as well as in *Globicephala*, according to the reconstruction adopted from the paper by Schreiber (1916), consists of the lamina transversalis anterior, the cartilago paraseptalis and the cartilago ductus nasopalatini. According to Eales (1950), the lateral cartilaginous bar of *Monodon* also consists of the three anlagen mentioned above.

In *Balaenoptera*, De Burlet (1914a) erroneously denoted as the cartilago paraseptalis only a small projection of the lateral structures, lying caudally of the connection place with the lamina transversalis anterior. In my opinion, this processus could at most correspond to a processus paraseptalis anterior. This projection, however, has never been found later in this form in any other specimen of baleen whales, either in *Megaptera* (Honigmann 1915, 1917) or in *Balaenoptera* and *Megaptera* (the present paper). Its existence must necessarily be questioned even because it was the very regio ethmoidalis of the specimen examined by De Burlet (1914a) that was damaged and thus the reconstruction remained incomplete (cf. also De Beer 1937). Although De Burlet (1914a) himself writes that the fact that unfortunately a part of the sections is missing in the middle of the rostrum does not substantially depreciate the reconstruction, its accuracy has nevertheless been doubted. It was possibly this questionable reconstruction by De Burlet (1914a) that later led Honigmann (1915, 1917), as well as De Beer (1937), to their wrong opinion that there is no cartilago paraseptalis in *Megaptera*. Later, Honigmann (1917) discussed this question again in great detail. Apparently influenced by the reconstruction of *Balaenoptera* by De Burlet (1914a), Honigmann (1917) believes to have distinguished, in an embryonic *Megaptera*, 114 mm in size, at least "small cartilaginous spines" as the remains of the cartilago paraseptalis. The proper paraseptal cartilage of *Balaenoptera* has been denoted by De Burlet (1914a) as the cartilago ductus nasopalatini, that of *Megaptera* by Honigmann (1917) and De Beer (1937) as the lamina transversalis anterior, the rostral tip of which passes into the cartilago ductus nasopalatini.

According to our earlier results (Klima and Van Bree 1985; Klima et al. 1986; Klima 1987a) as well as according to the observations described in the present paper, the cartilago paraseptalis of Cetacea can be characterized as follows (Figs. 38c,d, 47, 48, 49). The paraseptal cartilage of cetaceans is no complete thorough cartilago paraseptalis communis but only a transformed anterior part of that cartilage in the sense of a cartilago paraseptalis anterior, while the posterior part has been reduced to a processus paraseptalis posterior of varying size. The paraseptal cartilage is present in all cetaceans examined, both toothed and baleen ones, and it is the largest structure of the original nasal floor. Since in the cetaceans the Jacobson organ has not been preserved, not even in the form of a reduced anlage, the cartilago paraseptalis even forms no encompassing tube or supporting strut. Nor is there any fibula reuniens. The paraseptal cartilage is shaped like a simple ledge of varying breadth, flattened from lateral. The cartilage is intimate with the septum nasi, however without contacting it, and it reaches far anteriorly along the rostrum. The lower edge of the paraseptal cartilage is accompanied by the vomer all along its length, and the praemaxillare lies close to its lateral surface. In *Monodon* the rostral end of the cartilage is produced to form a semi-isolated cartilago ductus nasopalatini; in all other genera examined the two structures are continuously connected. The caudal part of the paraseptal cartilage continues dorsally in the lamina transversalis anterior, the boundary between these two structures being sometimes visible as a distinct gap (*Stenella*, *Delphinus*, *Lagenorhynchus*, *Phocoena*, *Monodon*), or as a fluent transition (*Globicephala*, *Physeter*, *Balaenoptera*, *Megaptera*). Below, the caudal end of the paraseptal cartilage may be formed into a more or less distinct processus paraseptalis anterior (*Stenella*, *Delphinus*, *Lagenorhynchus*, *Phocoena*).

6.3.4
Cartilago ductus nasopalatini

The cartilago ductus nasopalatini is an element of the anterior nasal floor, supporting the ductus nasopalatinus. It is absent from Monotremata and Marsupialia (De Beer and Fell 1936; Kuhn 1871) and is generally considered to be a neomorph structure of Placentalia (Terry 1917; Schliemann 1987, 1991). However according to the opinion of Kuhn (1971), it may not be a completely new formation of Eutheria, as it simply becomes isolated in the place where, in *Tachyglossus*, the lamina transversalis anterior, the cartilago paraseptalis and the cartilago palatina meet in front of the ductus

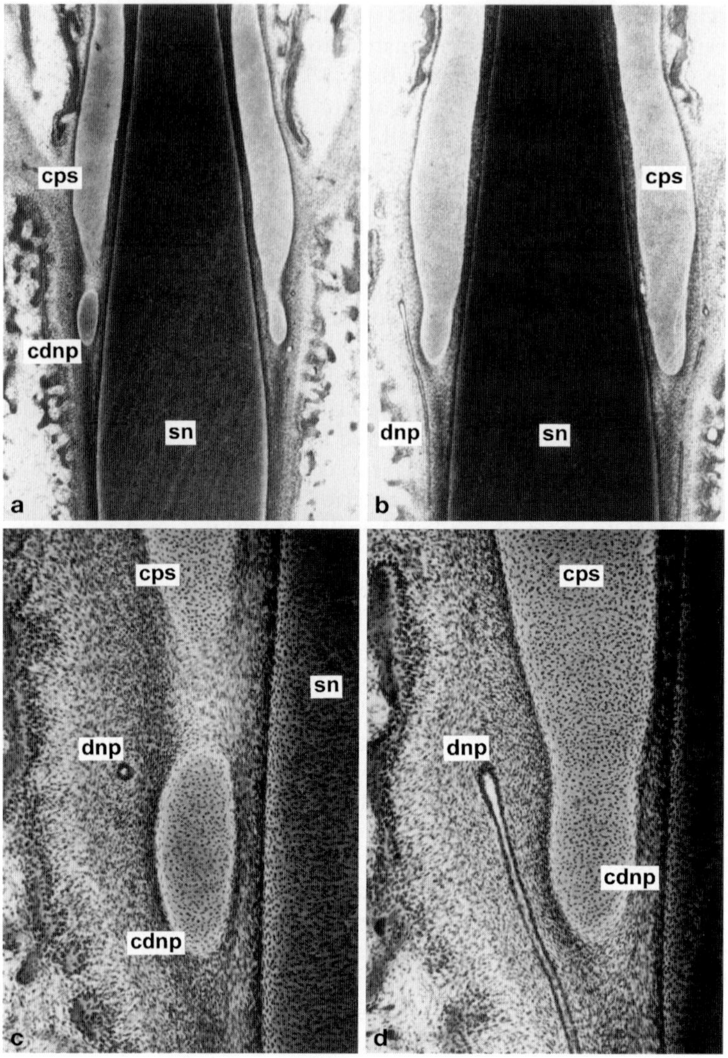

Fig. 49a-d. Cartilago ductus nasopalatini (*cdnp*) in *Monodon monoceros*, 137 mm TL. Horizontal sections in overall view (**a,b**) and in detail (**c,d**) from below to above.

nasopalatini. It does not attain its full development until the later stages of ontogeny (Schliemann 1987). It is rarely developed as a distinctly defined structure, as it mostly receives a close connection with other cartilaginous parts of the nasal floor. A simultaneous connection with the lamina transversalis anterior and the cartilago palatina is found in the cartilago ductus nasopalatini of *Myotis* (Chiroptera, Frick 1954), *Saimiri* and *Pithecia* (Primates, Maier 1981). In *Tupaia* (Scandentia, Zeller 1987) the cartilago ductus nasopalatini lies at the connection between the lamina transversalis anterior and the cartilago paraseptalis. The connection with the lamina transversalis anterior may be secondary, as suggested by the examinations of several Carnivora (Freund

1911; Ärnback Christie-Linde 1914; Terry 1917; Köhncke 1980, 1985; Sturm 1984; Schliemann 1987). The problem has been discussed in detail by Schliemann (1987) who writes the following on this point: "This cartilago ductus nasopalatini is connected to the lamina transversalis anterior by a vertically oriented piece of cartilage which provides mechanical support. This connection of the cartilago ductus nasopalatini to the lamina transversalis anterior ...presumably represents a derivated state."

Of all elements of the nasal floor, the cartilago paraseptalis shows the closest relation to the cartilago ductus nasopalatini. This relationship is fairly constant. Often the cartilago ductus nasopalatini appears only in the form of a hardly definable projection at the rostral end of the cartilago paraseptalis, with the ductus nasopalatinus being partly or wholly reduced. This form of a rostral projection of the cartilago paraseptalis is found in the cartilago ductus nasopalatini of some Insectivora (*Talpa*: Fischer 1901), Primates (*Pan*: Starck 1967; *Homo*: Reinbach 1963; Starck 1975), Carnivora-Fissipedia (*Cryptoprocta*, *Paradoxurus*, *Herpestes*, *Canis*, *Felis*: Schliemann 1987), Carnivora-Pinnipedia (*Leptonychotes*: Fawcett 1918), Sirenia (*Trichechus*, *Dugong*: Matthes 1912, 1921a), Hyracoidea (*Procavia*: Lindahl 1948), and Artiodactyla (*Dama*: Slabý 1990b), etc. Among the desmal bones of the embryonic skull, a close connection to the cartilago ductus nasopalatini is shown by the praemaxillare, which is intimate with the cartilage from the lateral and rostral sides (Frick 1954; Maier 1981; Schliemann 1987). An additional, somewhat looser connection to the vomer may develop at the lower posterior end of the cartilage.

The first note on the cartilago ductus nasopalatini of Cetacea is found in De Burlet (1914a) who describes, in *Balaenoptera*, a longitudinal cartilaginous bar lying lateral of the rostrum and representing, in its posterior part, the cartilaginous paraseptale; its anterior part should be considered to be the cartilago ductus nasopalatini. The place in which the lamina transversalis anterior is connected to the cartilaginous strut should be denoted as the boundary between these two parts. This interpretation is not tenable, however. The anlage described and also illustrated by De Burlet (1914a) as the cartilago ductus nasopalatini corresponds, in its most part, to the cartilago paraseptalis. In his later paper on *Lagenorhynchus*, De Burlet (1914b) was a little more careful in interpreting this anlage. Besides this lateral plate (i.e., the anlage of the cartilago paraseptalis Klima and Van Bree 1985), he found an additional cartilaginous part (i.e., the anlage of the processus lateralis ventralis Klima and Van Bree 1985), and he states that he hesitates to decide whether to denote, as the cartilago ductus nasopalatini, the end of this plate that protrudes towards the lower anterior region , or the free cartilaginous part lying in front and over this plate. Honigmann (1917), in his paper on *Megaptera*, supports the interpretation of De Burlet (1914a) in *Balaenoptera*, and he also denotes the whole anlage of the cartilago paraseptalis as the cartilago ductus nasopalatini. For his monograph, De Beer (1937) has adopted Honigmann's (1917) reconstruction of *Megaptera* with the same interpretation of the structures, but in the reconstructions of *Phocoena* and *Lagenorhynchus* after De Burlet (1913b, 1914b), and of *Globicephala* after Schreiber (1916), he has changed the interpretation so that he considers the largest part of the lateral plate to belong to the cartilago paraseptalis and denotes only its rostral tip as the cartilago ductus nasopalatini. This interpretation has also been supported by Eales (1950), yet it is in the very ethmoidal region that her reconstruction of the chondrocranium of *Monodon* shows several inaccuracies.

Of decisive importance for the precise localization of the cartilago ductus nasopalatini in Cetacea was the finding of the ductus nasopalatinus in *Monodon* (Klima and Van Bree 1985). At the rostral end of the paraseptal cartilage a deep constriction is observed, separating a small cartilaginous part from the bulk of the cartilage. Lateral of this constriction there is a narrow little tubulus lined with epithelium, rising vertically from below to above, then bending anteriorly and reaching rostrally along the partly isolated cartilaginous part (Fig. 49). However, this little tubulus is only rudimentary in character and it is obliterated at either end. Yet its morphology as well as its situation leave no doubt that it is the ductus nasopalatinus. Consequently, the partly isolated cartilage must be the cartilago ductus nasopalatini. This cartilaginous part is nowhere as distinctly defined as in *Monodon*, even though a suggestion of a division is apparent in most genera examined, at least in some of the stages. A temporary occurrence of the ductus nasopalatinus has also been demonstrated in some of the stages of *Stenella* and *Delphinus*. Thus, it is suspected that the anlage of the cartilago ductus nasopalatini remains preserved in the rostral tip of the cartilago paraseptalis, even in those cetaceans in which no definable cartilago ductus nasopalatini is present. This opinion is also supported by the fact that in all genera examined, a close connection of the cartilaginous structures to the desmal praemaxillare is found in this place.

7 The Nasal Structures of Adult Cetacea

In adult cetaceans (Figs. 50, 51) only some of the embryonic cartilaginous structures of the nasal skull are preserved as cartilage. The cartilaginous tissue either ossifies to form independent skull bones, or it becomes integrated into the bony tissue of the dermal bones, or it is dissolved and resorbed. These processes are readily understood if one realizes that most of the cartilaginous anlagen play only a transitory part during embryonic development. Capable of being built up quickly and simply, the cartilaginous tissue introduces the morphogenetic processes and paves the way for the bone tissue which is slower to build up and more complicated. The particular cartilaginous nasal elements are subject to the following modifications.

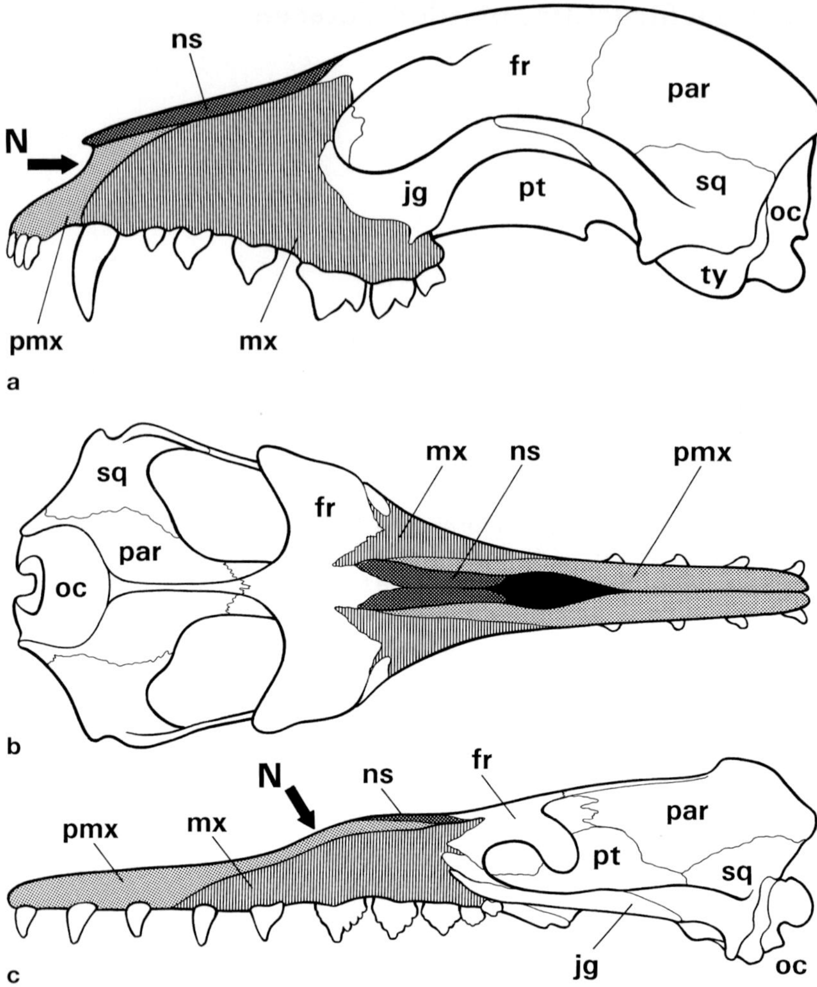

Fig. 50a-c. Translocation of the osseous skull elements from the generalized conditions in terrestrial mammals to the adapted skull of the archaeoceti. **a** The skull of a fox, *Vulpes vulpes*. **b,c** The skull of *Zygorhiza kochii* from the Upper Eocene. (**b,c** After Kellogg 1936)

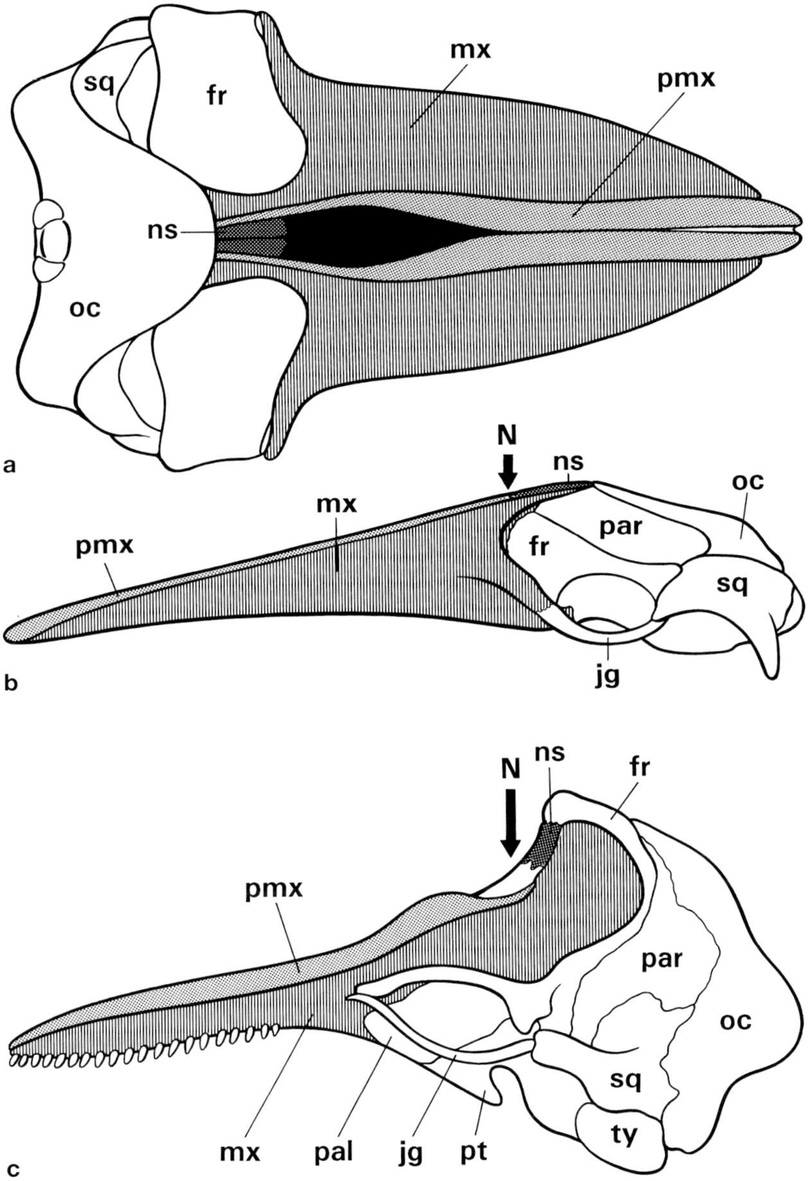

Fig. 51a-c. The osseous skull of adult Cetacea. **a,b** Fin whale, *Balaenoptera physalus*. **c** Common porpoise, *Phocoena phocoena*

7.1 The Ethmoid Complex

Through chondral ossification, the elements of the posterior side wall structures and the median structures develop into a single bony complex of the ethmoid bone.

Posteriorly, the ethmoid complex consists of a vertical transverse plate. It is slightly convex towards the anterior plane and forms a part of the posterior wall of the bony nasal passage. It is slightly concave towards the posterior plane and forms a part of the inner wall of the neurocranium. Above, the transverse plate is neighbors with the nasale, laterally with the maxillare and the palatinum, and below with the vomer. It shows precisely the same position as that of the posterior side wall structures in the embryonic nasal capsule. These structures, heavily reduced already during early morphogeny - at any rate in comparison with the nasal skull of most other mammals - have been still more reduced in the subsequent course of ontogeny. Their material has fused to form a simple, smooth bony plate in which any of the initial structures are no longer discernible. Thus, in this region we are up against a heavily regressive development of the nasal structures which is, beyond any doubt, the consequence of the loss of the sense of smell.

The anterior part of the ethmoid complex develops in a different way. It branches off the middle of the transverse plate in the form of a T-shaped, powerful and high bony lamella extending along the median plane anteriorly. With its lower edge, the lamella edges with the vomer with which it forms the septum nasi and which it accompanies rather far anteriorly as the bony snout (rostrum nasi osseum) until it passes into the cartilaginous snout (rostrum nasi cartilagineum). The whole bony lamella is the successor of the median structures of the embryonic nasal skull, in particular of the septum nasi and the rostrum nasi. However, these structures have been subject to a progressive rather than regressive development, especially as pertains to the anterior section, the rostrum, as they participate in the development of the typical long snout of the dolphins.

7.2 Cartilaginous Rostrum

The bony rostrum continues anteriorly into a cartilaginous process, the rostrum nasi cartilagineum. This is a long and narrow cartilaginous bar inserted between the bony elements of the upper jaw, i.e., the paired premaxillary bones, and the unpaired vomer bone, and accompanying these bone structures all along their length up to the snout tip. However, this bar is not preserved in macerated skulls. It only leaves a deep furrow along the mid-line on the upper surface of the upper jaw bones. It is present in all toothed whales and has been differently denoted as the “cartilago nasalis” (Kuzmin 1976), “cartilaginous vomer” (Clarke 1979a), “vomerine cartilago,” “rostrum” (Pilleri et al. 1983), “rostrum nasi,” “knorpeliges Rostrum” (Klima and Van Bree 1985) or simply as the “Oberkieferknorpel” (Klima 1990a). This cartilaginous structure develops from the anterior part of the embryonic nasal septum, i.e., from the anterior part of the septum nasi, which has been denoted as the embryonic rostrum nasi.

In most mammals the embryonic rostrum is later completely reduced or at least heavily so. In toothed whales, however, it is fairly well preserved even in adults. During embryonic development, this structure fulfills, among others, the task of introducing the growth and formation of the long anterior part of the head still before the bony elements appear (Klima and Van Bree 1985; Klima et al. 1986). In adult toothed whales this skeletal part probably plays a further part, namely that of the structure transmitting sounds produced for the purpose of echolocation. The still fiercely discussed problem of the transmitting and receiving mechanism of the so-called echolot or sonar system for remote orientation and communication in water (Reysenbach de Haan 1957; Fraser and Purves 1960; Kellogg 1961; Fleischer 1977, 1982; Purves and Pilleri 1983) seems to have been satisfactorily clarified in favor of a concept of the "exclusively laryngeal origin of the echolocation sounds in cetacea." Without going into details, let it be briefly noted in this place that the cartilaginous as well as bony structures in the upper jaw of the toothed whales, summarily denoted as the so-called rostrum, possess excellent sound-transmitting properties. This has been unequivocally demonstrated by exact structural examinations as well as bio-acoustical experiments. In this, even the rostral cartilage plays an important part. "The histological structure of the vomerine cartilago is very peculiar and finds no parallel in land mammals .... It is a fibrous cartilage with interwoven fiber bundles occupying an intermediate position between cartilage and bone and in which the collagen element predominates. The structural properties correlate very closely with the biophysical" (Pilleri et al. 1983).

The cartilaginous rostrum nasi is very well preserved even in adult Mysticeti. However, the purely bony skull specimens, in which the cartilaginous tissues have been dissolved through maceration, provide no true idea of the actual dimensions of this skull structure. The space of the mesorostral furrow (fossa mesorostralis), previously containing the cartilage, is covered for the most part by the praemaxillare and maxillae. Only the cross-sections through a non-macerated head (Lambertsen et al. 1989) allow one to learn the size of the share of the cartilaginous rostrum in the formation of the whole upper jaw (Fig. 52).

It is difficult to say, however, what is the task of the cartilaginous rostrum in baleen whales. It can hardly serve as a sound-transmitting structure, as there is no evidence of the presence of an echolocation system in these whales. However, since this structure shows a powerful development rather than reduction, it is bound to have a functional advantage. This advantage may possibly inhere in certain properties of the cartilaginous tissue. Evidently, being rich in water content, pressure-proof and, at the same time, elastic, the cartilaginous tissue possesses, in contrast to the bony tissue, certain properties that have proved to be particularly favorable for the adaptation of the skeleton to the aquatic life habits. Very probably, this is particularly true of large animal forms. Good examples thereof are provided by the cartilaginous skeleton of sharks, or the nasal roof cartilage in the anterior part of the sperm whale head (Klima 1990a).

Even the bony tissue shows significant modifications with increasing adaptation to the aquatic life habit. Thus, for example in the adaptation series of mammals from terrestrial (land mammals) to semiaquatic (Pinnipedia) to holaquatic (Cetacea), one can observe an increasing loss of massive and heavy compact bones and a distinct increase in the porous and lighter spongious ones (Klima 1994). The effect of gravity upon the body of land mammals necessarily leads to the development of massive skeleton. The bony skeletal parts mostly form a framework of tubular structures with a firm outer mantle of compact bone and with a loose core of spongious bone or even

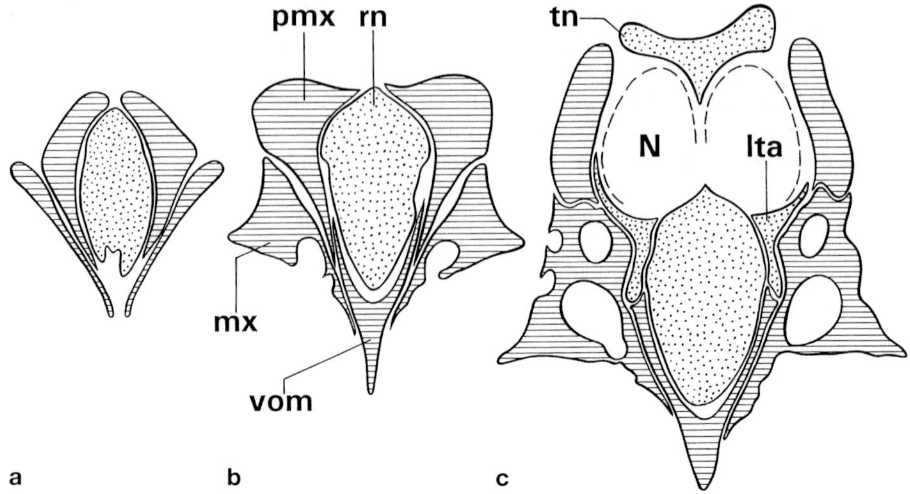

Fig. 52a-c. Cartilaginous rostrum nasi (*rn*) in the space of the fossa mesorostralis within the osseous skull of a subadult specimen of bowhead whale, *Balaena mysticetus*, of 7.5 m body length. Transverse sections from anterior to posterior (Altered from Lambertsen et al. 1989)

boneless marrow cavity. The change to aquatic life habits removes the one-sided burden caused by gravity. The hydrostatic pressure as well as the lifting power of water affect the body in a more even and alleviating manner (Hertel 1963). The massive compact bone is no longer necessary, and a fine-mesh spongious bone fills the bony parts of the skeleton completely and uniformly (Felts and Spurrel 1965; Felts 1966; Klima 1994).

Even the cartilaginous tissue could possibly possess similar favorable hydrostatic properties. This would explain why, in contrast to land mammals, the cetaceans possess such well-preserved cartilaginous parts in their skull. This might be the true and primary cause of the preservation of a cartilaginous rostrum nasi, not only in the baleen whales but even in the toothed ones. Its function of a sound-transmitting structure serving the sonar system of the toothed whales could then have developed as an additional secondary task. If, however, even the baleen whales should initially have possessed a sonar system that was reduced later on (Heyning and Mead 1990), then their cartilaginous rostrum should be likewise considered to be a remainder of that system.

7.3 Cartilages of the Nostril

Around the nostril there are small cartilaginous structures (the cartilagines aperturae nasi), which in several Cetacea, have been found to be parts of the mechanism closing the nostril. They show considerable interspecific variability as regards their number, size, form and arrangement. Very little is known of these structures. In the literature, they have been mostly mentioned only in the form of marginal notes to other topics (e.g., Von Baer 1826; Kükenthal 1893; Gruhl 1911). Apparently, they serve the purpose of strengthening the muscles closing the nostril. During embryonic development, they develop from the material of the anterior nasal capsule, the cupula nasi anterior, as follows, e.g., from our investigations in *Phocoena* (Klima and Van Bree 1985).

More particulars have been ascertained on the nostril cartilages in the sperm whale *Physeter catodon* (Behrmann and Klima 1985; Klima et al. 1986). The left embryonic nasal capsule, the cupula nasi anterior sinistra, is preserved in the form of several massive cartilaginous parts that form together a ring around the left nostril. The diameter of the whole cartilaginous ring is about 80 cm, the individual cartilaginous parts are up to 20 cm in size. The cartilages are inserted in a thick ligamentous tissue with which they form the support for the nostril muscles. In contrast to the other toothed whales, in the sperm whale the nostril is not situated on the vertex of the skull but on the upper tip of the giant, bulgy anterior part of the head that contains the well-known spermaceti organ. The two very long nasal passages, asymmetrical and of different size, run at different heights obliquely through the tissue of the spermaceti organ. Only the left nasal passage serves the direct air supply and opens to the surface as the nostril. The right, somewhat transformed nasal passage has no relation to the nostril. The right nasal capsule, the cupula nasi anterior dextra, still present in the early embryonic stages, is no longer present in the adult sperm whale.

The histological structure of the nostril cartilage of the sperm whale is interesting (Klima 1990a). In the embryos and fetuses, these structures consist of a hyaline cartilage; in the adults it is the same special form of the elastic cartilage as ascertained in the great nasal roof cartilage (see below), except that it contains a larger amount of cells and fewer tissue parts distinctly capable of damping down pressure.

7.4 Nasal Roof Cartilage in the Sperm Whale *Physeter catodon*

The embryonic nasal skull of the sperm whale shows several peculiarities. The most conspicuous among them is the presence of a powerful tectum nasi which, together with the anlagen of the cupulae nasi anteriores, spreads obliquely throughout the anterior part of the head like a unified bar (Fig. 53). The sperm whale is the only toothed whale showing such well-developed tectum. In all other embryonic toothed whales examined so far (*Monodon*, *Phocoena*, *Globicephala*, *Lagenorhynchus*, *Stenella*) the tectum nasi is almost entirely reduced (Boenninghaus 1903; De Burlet 1913a,b, 1914b, 1916; Schreiber 1916; De Beer 1937; Eales 1950; Klima 1987a; Klima and Van Bree 1985). In the embryonic development of the sperm whale, the tectum nasi plays an important part, being the leading momentum for the anterior head, as demonstrated above. At the same time, it serves to support the nasal passages, an increasing connection being observable between it and the left nasal passage during ontogeny.

Since we have ascertained this cartilaginous anlage even in larger fetal stages in an undiminished development, the anticipation was that at least a rudimentary tectum would have to be present in adult sperm whales as well. This was confirmed by the examination of a male sperm whale, 18 m in length and weighing 57 tons, that stranded near Bremerhaven in 1984 (Behrmann 1985; Klima and Behrmann 1985; Behrmann and Klima 1985). In that whale, the tectum nasi passes, like a flat cartilaginous band, from the floor of the bony nasal skull diagonally through the whole soft anterior part of the head, accompanying the left nasal passage up to the nostril.

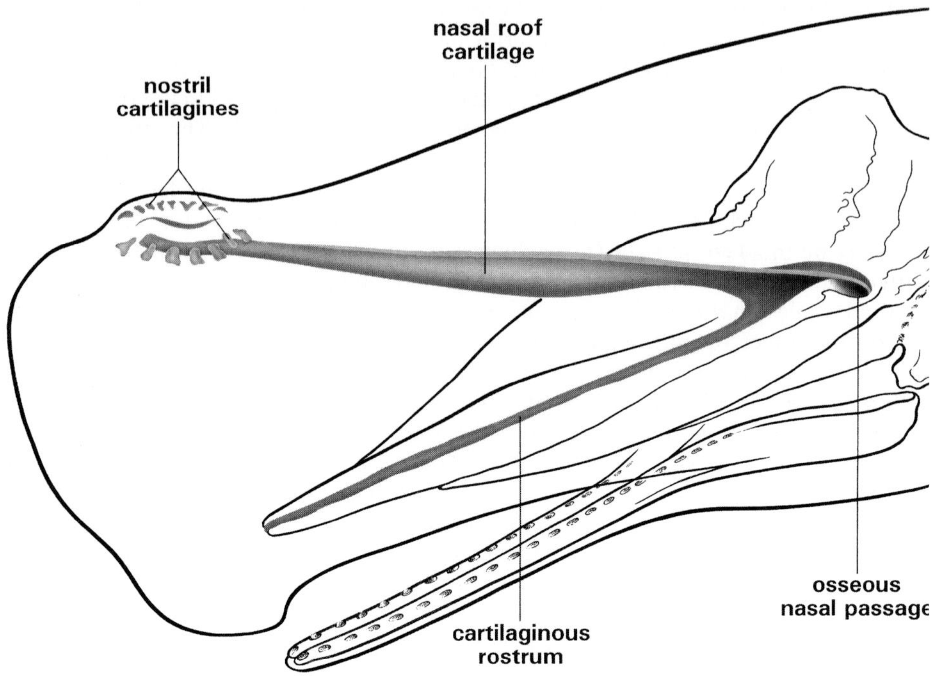

Fig. 53. Outline drawing of the head of an adult sperm whale, *Physeter catodon*, with the position of the nasal roof cartilage

It is hard to believe that such a conspicuous structure should remain unknown until 1985. None of the authors examining the head of the sperm whale (Pouchet and Beauregard 1885, 1889; Beddard 1915, 1919; Raven and Gregory 1933; Schenkkan and Purves 1973; Kuzmin 1976; Clarke 1978a,b, 1979a,b) found a cartilage in the so-called soft parts of the giant fore part of the sperm whale head. No mention is found of the presence of a cartilaginous tissue in the fore part of the sperm whale head in the literature specially concerned with the sperm whale (Matthews 1938; Berzin 1971; Norris and Harvey 1972; Arvy and Pilleri 1983), nor in the most important standard works on whales (Kükenthal 1983; Kellogg 1928; Slijper 1936, 1962; Walker 1964; Watson 1981; etc.). Clarke (1978a) is the only author who mentions a "cartilaginous vomer," which he had found in a fetus at the level of the nasofrontal sac, i.e., rather far caudally, immediately in front of the bony neurocranium. However, this author found no cartilage in the rostral part of the fetal head. Neither did Clarke (1978a,b, 1979a,b) find any cartilaginous structures in the anterior head part of an adult sperm whale examined by him, his observations being the most thorough ones published so far. Clarke (1978b) only mentions "the tough fibrous lining of the narrow left passage."

During embryonic development, the nasal cartilage plays an important part as a structure that leads the growth from the inferior posterior diagonally towards the rostral superior region and is responsible for the formation of the soft fore part of the head (Klima et al. 1986). In this way the head develops its typical "sperm whale shape."

The persistence of this structure even in the adult sperm whale is also bound to have a functional significance.

A function of the nasal roof cartilage in the sense of a true skeletal strut in the soft fore part of the head is certainly out of question. For this, the cartilaginous part is simply too weak and the soft tissue too powerful. The immense sperm whale head could not be supported by a single skeletal part at all. Only a relatively small part of the head possesses a bony scaffolding which has been degraded to an unimportant marginal construction. It consists of slender jaws and a compact neurocranium. According to Clarke (1979b), the bony parts account for only about 12% of the head mass, the remaining 88% consisting of soft parts. They include the far more massive part of the head, resembling the stream-lined bow of a ship and consisting of ligamentous tissue rich in oil and of muscles (Pouchet and Beauregard 1885, 1889; Beddard 1915, 1919; Raven and Gregory 1933; Matthews 1928; Berzin 1971; Norris and Harvey 1972; Schenkkan and Purves 1973; Kuzmin 1976; Clarke 1978a,b, 1979a,b; Arvy and Pilleri 1983). There is also the so-called spermaceti organ of the sperm whale, containing an oily liquid, the spermaceti, which, according to Clarke (1979c), enables the whale to counterbalance the lifting powers when diving down to depths around 1000 m (in extreme cases as deep as around 3000 m). This huge soft mass, which may weigh over 20 tons and amount to about a third of the total body weight (Clarke 1979b), is conceivable as a functional structure only when in water. Only the lifting powers prevailing in water enable it to retain its shape and function. Ashore it would collapse due to its own weight, as could be ascertained during preparation (Behrmann 1985). In view of the above, then, the function for the nasal roof cartilage as a skeletal strut is quite insignificant.

According to histological examinations, the nasal roof cartilage of the fetal sperm whale consists of embryonic hyaline cartilage. This kind is especially suitable for the developmental processes and a rapid growth. In adult sperm whales (Fig. 54), the nasal roof cartilage consists of a special form of elastic cartilaginous tissue closely resembling the hyaline cartilage (Klima 1990a). The arrangement of cells, their territories as well as the extracellular substance correspond to the hyaline cartilaginous tissue. This component represents an adaptation to pressure stress. The occurrence and arrangement of elastic fibers correspond to the elastic cartilaginous tissue. This component represents an adaptation to deformation stress. Both pressure and deformation are obviously the most intensive requirements to which the nasal roof cartilage is exposed. I can see the function of this cartilaginous structure in that, as a pressure-resistant skeletal part accompanying the left nasal passage all along its length through the massive soft fore part of the head, it secures the only direct air passage. Besides, bundles of striate muscle fibers are anchored in the perichondrium of the nasal roof cartilage. The task of this finely interwoven muscular system can be seen in that it controls the fine movements of the contracting and dilating nostril.

Besides, it is quite possible that, like the cartilaginous rostrum, also the nasal roof cartilage functions as a sound-transmitting skeletal part serving the sonar system. In fact, the two elements are continuously connected mutually, thus forming a skeletal unit which probably also shows uniform biophysical properties. In this connection, however, I wish to point out a possibly important difference in the histological structure of the two parts. Pilleri et al. (1983) found no elastic fibers in the rostral cartilage of the toothed whales: "Elastic elements are completely absent." In contrast to that, the nasal roof cartilage of an adult sperm whale consists of an elastic

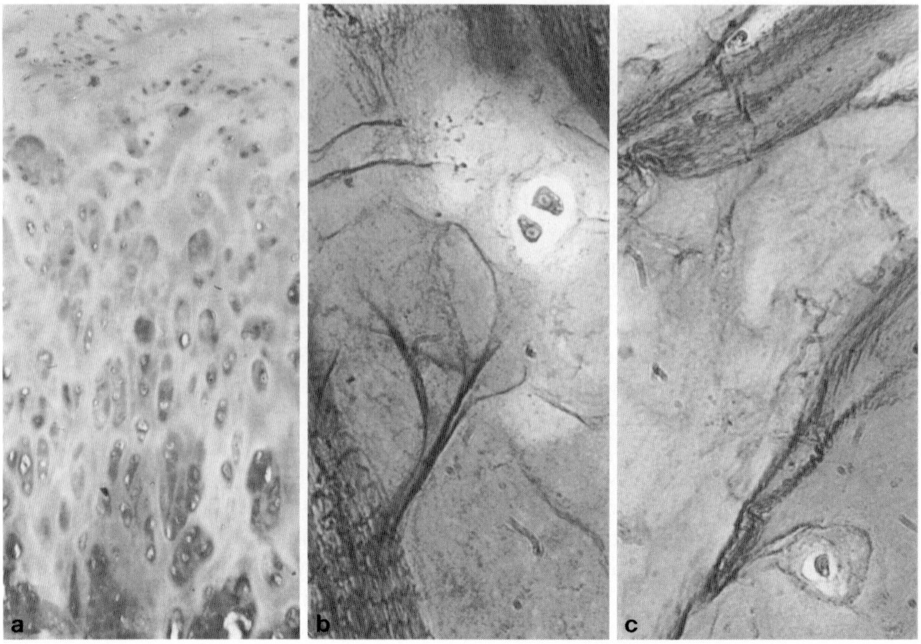

Fig. 54a-c. Cartilaginous tissue of the nasal roof in an adult sperm whale, *Physeter catodon*, of 18 m body length. **a** The arrangement of cells and their territories in overall view, stained with azan, approximate scale 60:1. **b,c** Cells and elastic fibers in detail, stained with resorcin, approximate scale 450:1

cartilaginous tissue. To what extent this fact may affect the acoustic properties of the structure, I am unable to decide.

7.5 Pararostral Bones, or the so-called Meckelian Ossicles

As described in detail above, the anterior side wall structures participate, during embryonic development, in the formation of the typical cetacean rostrum in a decisive way. It is unthinkable that such powerfully developed, as well as functionally important structures, should be completely reduced during subsequent ontogeny. Their material is incorporated in the bony elements of the rostrum as a whole. It may either merge into them or remain preserved as an independent skeletal element.

Attempting to define, in the skull of adult cetaceans, the region in which the anterior side wall structures are situated during embryogeny, it appears that these structures are bound to be rather concealed (Fig. 55). For the most part they are intimate with the medial surface of the posterior section of the praemaxillare. Still more to the inferior, they neighbor with the maxillare and the vomer, posteriorly with the ethmoidale. Only their superior posterior tip, situated above the praemaxillare, may be seen in the outer view of the skull. It is this very place in which the so-called Meckelian ossicles are occasionally found (Fig. 56).

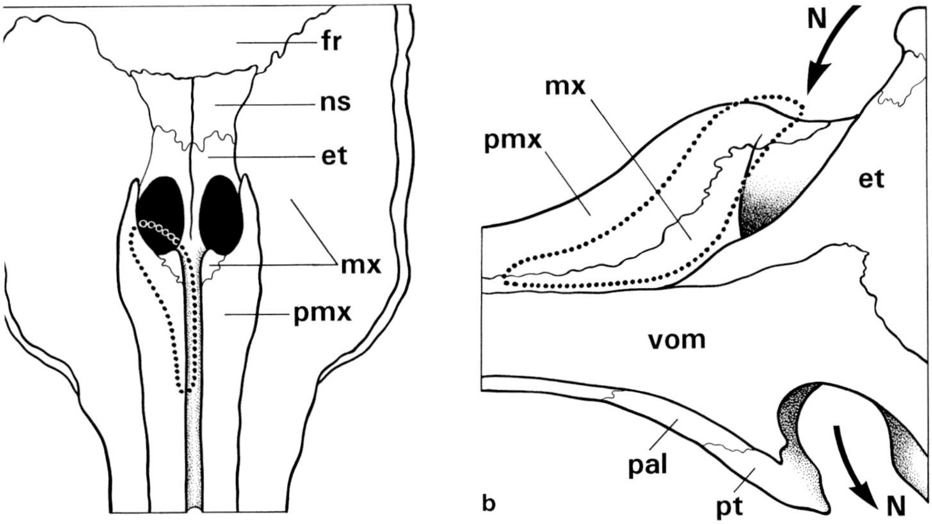

Fig. 55 a, b. The topographic position of embryonic anterior side wall structures (*dotted line*) indicated in the nasal skull of an adult *Phocoena phocoena* in dorsal view (**a**) and medially (**b**)

In the literature, these structures have been described for the first time by Meckel (1825) in the skull of *Phocoena*. However, they had already been observed and illustrated earlier than that. Probably the earliest illustration of these structures is found in the paper by Tyson (1680), as recently discovered by Cave (1987). These structures have been generally thought to be vestigial maxilloturbinals. This interpretation has been erroneously attributed to Von Baer (1826), although in fact it comes from Mayer (1835), as recently corrected by Cave (1987). Certainly, there is no concrete evidence supporting the derivation of the Meckelian ossicles from the maxilloturbinals. Rather, this hypothesis has been based on solely theoretical considerations. Even Kükenthal (1893) and Boenninghaus (1903), who were the first to examine a few embryos for this purpose, homologized the Meckelian ossicles with the embryonic maxilloturbinals. Since that time, no more attention has been paid to this question until recently.

On the basis of our examinations of an extensive and continuous developmental series of embryonic *Phocoena*, we have been able to tackle this question anew (Klima and Van Bree 1985). Our investigations have provided unequivocal evidence showing that the maxilloturbinals, being a part of the embryonic nasal wall, the paries nasi, are considerably reduced and their rudiments are shifted not towards the anterior but to the posterior edge of the nasal passages. Besides, they show a firm connection with the maxillary bones situated posteriorly. On these grounds the maxilloturbinals are out of question as the possible initial structure of the Meckelian ossicles. The Meckelian ossicles develop from the elements of the embryonic nasal floor, the solum nasi. They are "... homologous with the lamina transversalis anterior and also contain the material of the cartilago paraseptalis" (Klima and Van Bree 1985).

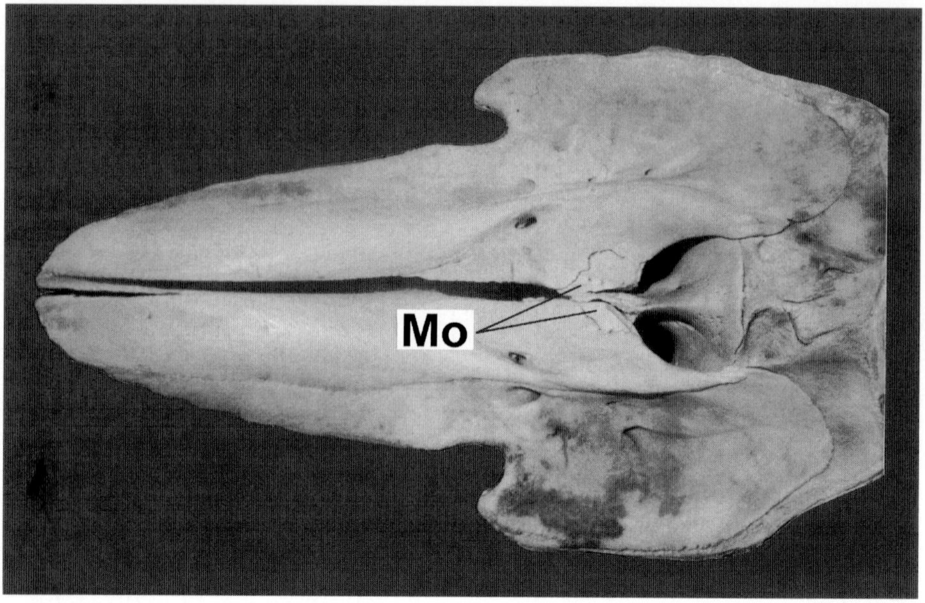

Fig. 56. Ossa pararostralia or the so-called Meckelian ossicles (*Mo*) in the skull of an adult *Pseudorca crassidens*. Photograph by G. Behrmann

Moreover, we have found that these embryonic structures of toothed whales have increased in size and that they have taken up a new functional significance. The above facts have lead us to the following prediction: The incidental occurrence of these elements in form of rudimentary skeletal parts in the nasal skull of adult odontocetes would thus be quite understandable, and it is to be expected that they may be found in several forms" (Klima and Van Bree 1985). In an impressive way, this prediction has been confirmed by the findings of Cave (1987): The ossicle was ... observed in situ in crania of the genera *Pontoporia, Delphinaptera, Monodon, Delphinus, Stenella, Tursiops, Lagenorhynchus, Pseudorca, Orcaella, Feresa, Phocaenoides* and *Phocoena* and traces of it were encountered in *Kogia*. It was found most frequently in crania of *Phocoena, Monodon, Delphinaptera* and *Lagenorhynchus*." Furthermore, Cave (1987) writes: More recently the Meckelian ossicles has been unequivocally depicted in *Phocoena, Grampus* and *Monodon* by Van Beneden and Gervais (18681880), in *Pseudorca* by Flower (1866), in *Phocoena* by Reynolds (1897, 1913), in *Tursiops, Platanista, Sousa, Neomeris* and *Delphinus* by Pilleri and Gihr (1969, 1971, 1972a,b), in *Pseudorca* by Purves and Pilleri (1978) and in *Cephalorhynchus* by Van Bree (1979), but in no instance is reference made in accompanying text or legend to the ossicle's presence."

Even if the Meckelian ossicles are not present as independent structures, their contours can be seen in the bony sutures in the praemaxillare. In another paper of his, Cave (1988) writes the following: Such suture persistence is sporadic in certain odontocete genera. Its cause remains obscure, but it plainly represents an arrested stage of rostrum development. Examples of such arrest have been presently observed in specimens of *Platanista, Pontoporia, Delphinapterus, Kogia, Monodon, Delphinus,*

Lagenorhynchus, Cephalorhynchus, Stenella, Orcinus, Pseudorca, Orcaella, Globicephala, Phocaenoides, Phocoena, Neophocaena, Feresa, Tursiops, Sousa, Grampus and *Tursiogrampus* forms in which, ideally, the ossified ossiculum narium unites with the praemaxilla without external trace remaining of its original anatomical independence."

Apparently, such structures are also involved in the skull of the adult *Physeter* examined by us (Behrmann and Klima 1985). We have denoted them as the vomer (Fig. 5, Behrmann and Klima 1985), as they show a close connection to the vomer, yet by their position and shape, they must be the Meckelian ossicles which are defined from the vomer and the praemaxillare by suturae. Like the whole skull of *Physeter*, even the Meckelian ossicles show extreme asymmetry. The skeletal part of the large left nasal passage is massive, long and projecting far rostrally; on the contrary, that of the smaller right nasal passage is slender, short and posited more towards the occipital direction. We have already ascertained the same asymmetry in the shape and position of the lamina transversalis anterior and the cartilago paraseptalis in embryonic *Physeter* (Klima et al. 1986; Klima 1987a). Similar structures in the skulls of adult sperm whales have also been depicted several times (Van Beneden and Gervais 1868-1880; Arvy and Pilleri 1983), however without any mention being made of their identity.

So far, the Meckelian ossicles have not been ascertained in Mysticeti. The investigations of Cave (1988) in this line have been unsuccessful. Nevertheless, the ossicles described and depicted by Eschricht and Reinhardt (1861) in the skull of *Balaena* as the first, second and third turbinate bones" could clearly be denoted as remainders of the embryonic anterior side wall structures and, consequently, the Meckelian ossicles. I consider Caves (1988) opinion on this finding, [i]t is possible that none of these structures was intended to represent a maxilloturbinale and that, in fact, all three are ethmoturbinates," to be incorrect. Naturally, no maxilloturbinalia are involved but certainly these structures cannot be identified with the ethmoturbinalia either, as indicated by my findings in the skulls of embryonic *Balaenoptera* and *Megaptera* in the present contribution.

In *Balaenoptera siberi*, a late Miocene baleen whale recently described by Pilleri (1989) from Peru, the fossils of which I have also examined, there are isolated ossicles immediately beside the caudal tips of the praemaxillaria, considered to be the Meckelian ossicles by Pilleri (1989). Although the displacement of these structures behind the nasalia (Fig. 57, Mo1) is in fact at variance with the typical situation of the Meckelian ossicles, they may nevertheless be considered to be a part of them. As a matter of fact, they form the accompanying structures of the praemaxillaria, which in the Mysticeti have been displaced far caudally up to the base of the nasalia. As a matter of course, I consider even the much larger and somewhat irregularly formed structures situated more rostrally (Fig. 57, Mo2) to be rudiments of the embryonic anterior side wall structures. As a whole, they form a slender cone inserted between the praemaxillare and the vomer and protruding far rostrally along the mesorostral fossa. As a whole, these structures even show much the same form as the structures found in the Odontoceti in the same place. Thus, we have actually the Meckelian ossicles even in the Mysticeti. In my opinion, little attention has been paid to this question so far, and more detailed investigations will sooner or later reveal the presence of the Meckelian ossicles even in the Mysticeti. In this point, the embryonic relationships show no major differences between the Odontoceti and the Mysticeti, as demonstrated in the present paper.

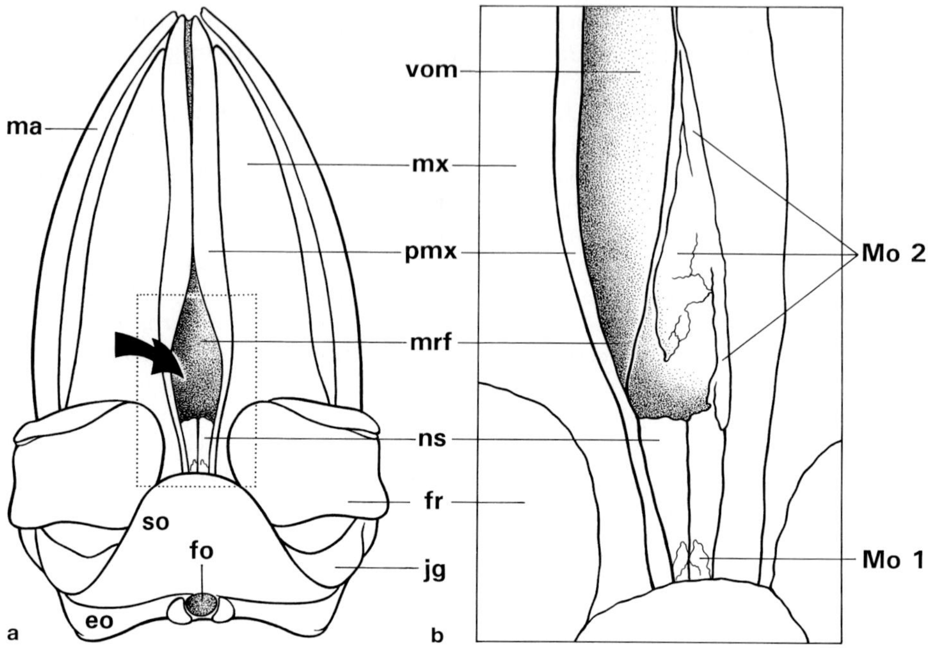

Fig. 57a-b. Ossa pararostralia or the so-called Meckelian ossicles (*Mo1, Mo2*) in the nasal skull of *Balaenoptera siberi*. **a** Complete view; the *dotted line* indicates the detailed view (**b**) into the inside of the mesorostral fossa (*mrf*). For details see text

Many of the ossicles found by Cave (1987, 1988) and Praderi and Cave (1994) show a surprising agreement in both shape and topography with the embryonic structures found by us (Figs. 58, 59), thus indirectly confirming our conception of the origin of the Meckelian ossicles from the lamina transversalis anterior and the cartilago paraseptalis. Nevertheless, in ignorance of our results, Cave (1987) advocates the theory, falsified by us, of the Meckelian ossicle originating from the maxilloturbinals. He writes: "Dissection of both fresh and macerated odontocete material demonstrates that the Meckelian ossicle is simply the externally visible dorsal portion of an ossiculum narium, the odontocete maxilloturbinate." In another place, Cave (1987) denotes these ossicles as "a vestigial maxilloturbinate."

In this place, let us emphasize again that the findings presented in the paper mentioned above (Klima and Van Bree 1985), as well as the arguments presented in our subsequent papers in this line (Klima 1987a; Klima and Van Bree 1990), unequivocally refute the origin of the Meckelian ossicles from the maxilloturbinals. In this sense, even the present paper provides further evidence, above all, by the results obtained from the developmental series of embryonic *Stenella* and *Delphinus*.

In my opinion, the Meckelian ossicles are in no case rudimentary structures, definitely not in the sense of vestiges of earlier structures that now and then occur in the recent forms, on the whole being subject to regressive development leading to their

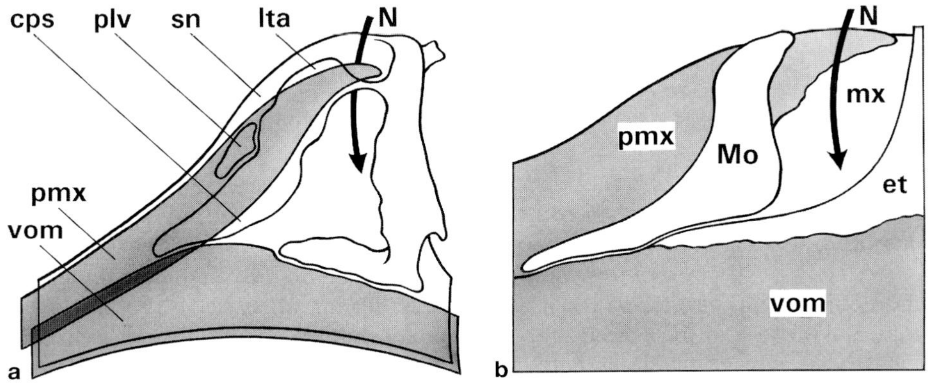

Fig. 58a-b. Nasal skull of *Lagenorhynchus* in the lateral projection. **a** Cartilaginous nasal capsule and dermal bones praemaxillare (*pmx*) and vomer (*vom*) of an embryo (altered from Klima and Van Bree 1985). **b** Schematic view of the bony nasal skull of an adult specimen. The anterior side wall structures of the embryonic nasal skull (*cps, plv, lta*) show the same shape and the same topographic position as the Meckelian ossicles (*Mo*) in the adult specimen (altered from Cave 1987)

complete disappearance. Quite the contrary. I consider them to be structures undergoing progressive development and being on the way towards stabilization in the evolutionary historical sense. This opinion is quite in accordance with the general mode of morphogeny as estimated by Slabý (1986):

"Old ancestral reducing and regressing characters can be utilized as building material, since evolution is always short of material for the development of new adaptive structures and for structural development it has to make use of what happens to be available. The persistent presence in the genome of the structural and organizational genes relevant for the given reducing structures and their inclusion in the morphogenetic program by the part of the genome controlling morphogenesis are an essential prerequisite, however. Changes in this morphogenetic program during evolution take place in the early stages of morphogenesis, in the blastema stage, when new differentiation of the given structure, under the above conditions, is still possible."

The embryonic structures of the nasal floor, only moderately developed in other mammals, in particular the lamina transversalis anterior and the cartilago paraseptalis, to which the Meckelian ossicles are to be related, have experienced a powerful development in the Cetacea, as they have taken up new tasks. Instead of participating in the formation of the heavily reduced nasal skeleton, they have become a decisive growth factor of the characteristic cetacean rostrum. One should bear in mind that they are among the regularly occurring skeletal parts of the rostrum. They will also have to be mentioned when enumerating the individual components of the whole rostrum of the cetacean skull, i. e., together with the praemaxillare, maxillare, vomer, and rostrum.

Instead of being vaguely denoted as Meckelian ossicles, or quite erroneously as the ossiculum narium, the odontocete maxilloturbinate (Cave 1987), or the cetacean os maxilloturbinare (Cave 1988), these structures should be better called the pararostral

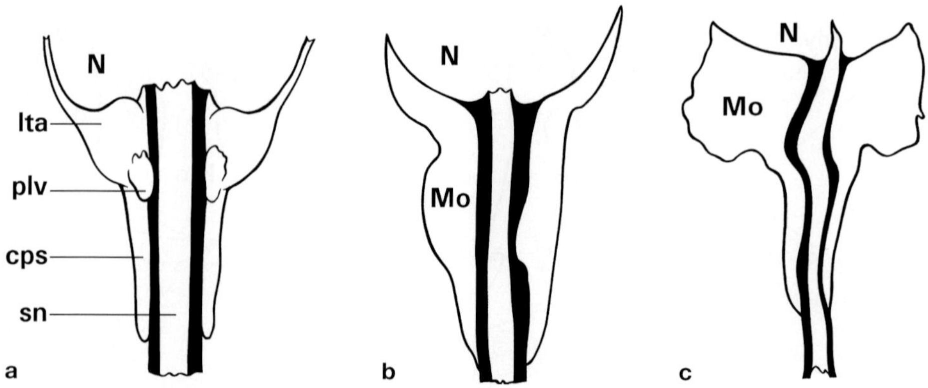

Fig. 59a-c. Nasal skull of *Stenella* in dorsal view. Comparison of the anterior side wall structures (*lta*, *plv*, *cps*) of the cartilaginous nasal capsule in an embryo (**a**) with the Meckelian ossicles (*Mo*) in two adult specimens (**b,c**) (b and c altered from Cave 1987)

bones, or the ossa pararostralia. Such a denomination clearly indicates that these structures have lost their pertinence to the nasal skeleton and that they have become, both topographically and functionally, a part of the entire cetacean rostrum. Besides, avoiding the diminutive ossicles" clearly indicates that they are definitely no tiny bone rudiments but, in many cases, considerably large skull bones.

8 Evolutional and Functional Relationships

8.1 Position of the Nostrils and the Mode of Surfacing in Cetaceans

The way in which the nostrils have been translocated during the evolution of cetaceans can be traced in fossil skull material as well as in the course of the embryonic development of their skull.

In terrestrial ancestors of the cetaceans the nostrils were still situated at the anterior tip of the skull in the position typical of most land mammals. With advancing adaptation to the aquatic habitats and with the obviously increasing speed of swimming, the nostrils shift upwards to the dorsal side of the nose and are then translocated more and more posteriorly (Fig. 60). At the same time, the splanchnocranium with the upper jaw is produced to form a narrow rostrum

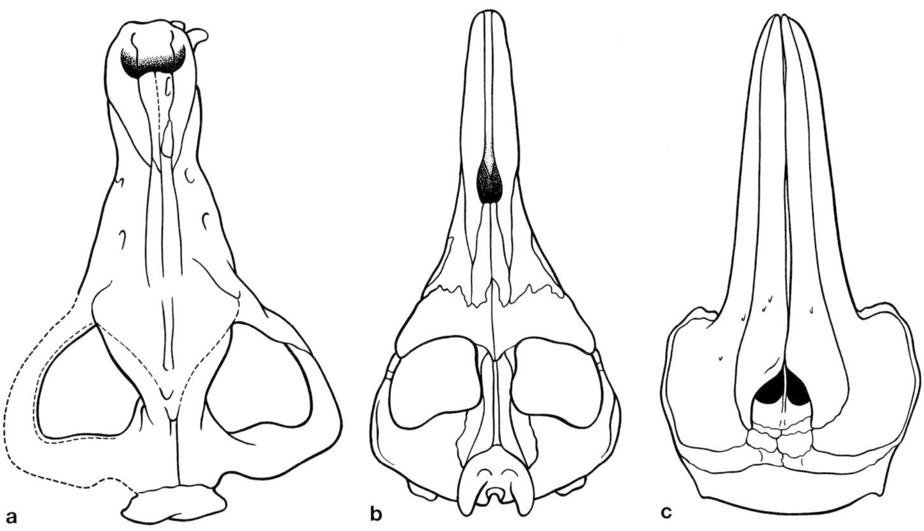

Fig. 60a-c. Translocation of nostrils in cetaceans in the course of evolution, as demonstrated by the different position of the external nasal apertures (*black*) in dorsal view of the skull. **a** *Andrewsarchus*, Mesonychidae, an ancestor of cetaceans from the Eocene. **b** *Protocetus*, Archaeoceti, ancient form of cetaceans from the Eocene. **c** *Tursiops*, Odontoceti, a recent cetacean (From various sources, redrawn by H. Schneeberger)

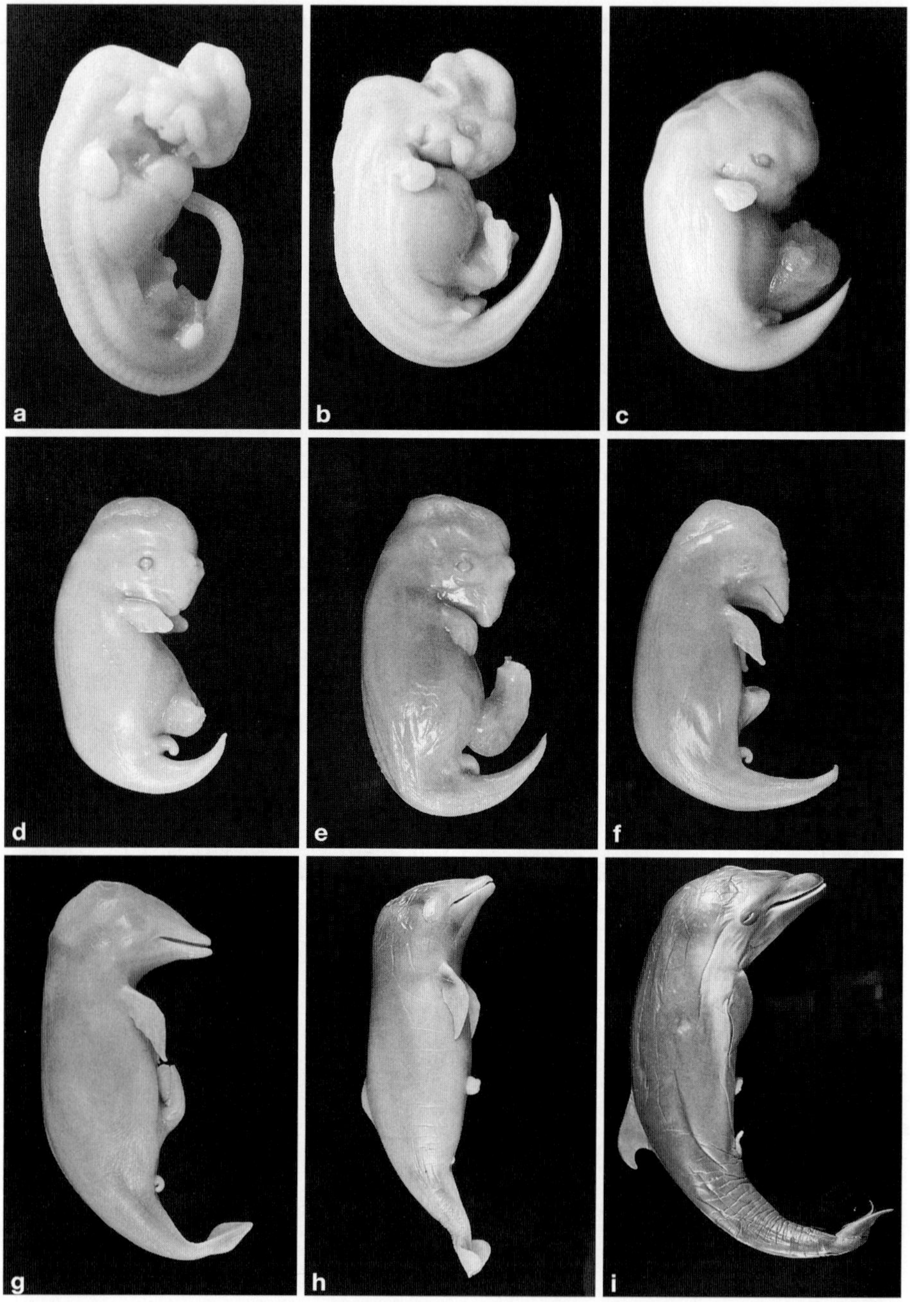

Fig. 61a-i. Translocation of nostrils from the snout tip to the vertex of the head during embryonic development of the spotted dolphin, *Stenella attenuata*. Size of the embryos in mm TL: **a** 25; **b** 38; **c** 43; **d** 57; **e** 67; **f** 76; **g** 91; **h** 153; **i** 477. Material from the collections of the Dr. Senckenbergische Anatomie, Frankfurt/Main (Photographs by B. Schildger)

Fig. 62a-f. The nostrils of cetaceans are always situated at the highest point of their whole body. They are paired in the baleen whales: **a** Blue whale, *Balaenoptera musculus*; **b,c** fin whale, *Balaenoptera physalus*; **d** Bryde's whale, *Balaenoptera edeni*. **e,f** The single nostril of the sperm whale, *Physeter catodon*, is situated asymmetrically on the left front side of the enormous boxlike head (Photographs from Cousteau 1988, Cox 1990 and Martin 1992)

protruding far anteriorly. Also, the neurocranium becomes shorter, the frontale and the parietale are regularly pushed back and the whole neurocranium becomes flattened from anterior towards posterior. Thus, the nostrils pointing upwards are positioned on top of the skull and the nasal passages pass from there vertically down between the splanchnocranium and the neurocranium. With minor differences, these changes pertain equally to both suborders of Cetacea, i.e., the toothed whales (Odontoceti) as

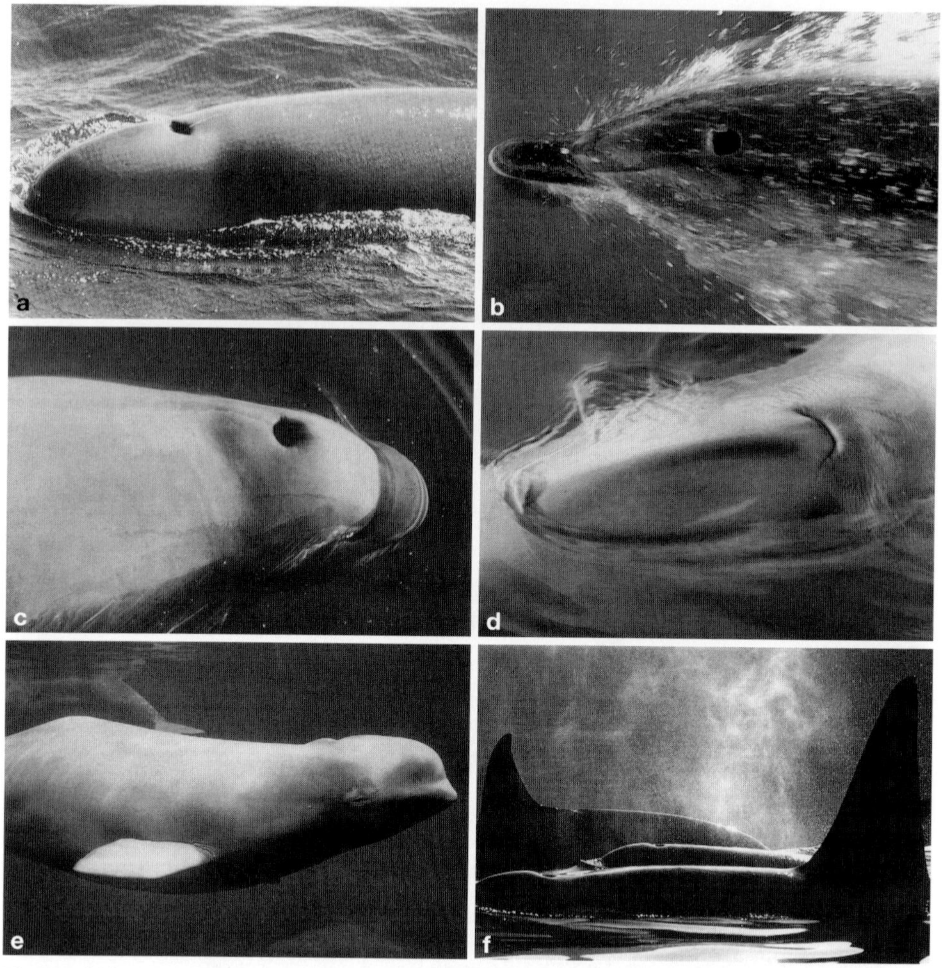

Fig. 63a-f. The nostrils in the toothed whales have fused to form an unpaired blowhole on the vertex of the head. **a** Northern bottlenose whale, *Hyperoodon ampullatus*; **b** spotted dolphin, *Stenella attenuata*; **c-e** beluga whale, *Delphinapterus leucas*; **f** killer whale, *Orcinus orca* (photographs from Cox 1990, Gewalt 1976 and Martin 1992)

well as the baleen whales (Mysticeti) (cf. Kellogg 1928; Slijper 1962; Gaskin 1976; Oelschläger 1978, 1987, 1990; Moore 1981).

A kind of repetition of these evolutional processes can also be observed during the ontogeny of cetaceans, as already pointed out in the literature (e.g., De Burlet 1916; De Beer 1937; Klima and Van Bree 1985; Klima 1987a; etc.) and as documented by the results of the present study. At the earliest embryonic stages, the nostrils are still positioned at the tip of the snout, as in land mammals; at older stages they are successively shifted towards the top of the head. At the same time, a long rostrum with narrow jaws develops (Fig. 61).

The nostrils, translocated to the middle of the vertex, lie in the highest point of the cetacean body. This is the first point to emerge from the surface of water when the cetacean surfaces, and it is the only place on its body that sticks out of water when the resting cetacean lies motionless, close beneath the water surface. Surfacing in order to breathe out and in, cetaceans need not at all interrupt the continuous course of their natural swimming movements. Moving in the typical mammalian way by undulating their body in the dorsoventral direction, they only need to emerge a minute fraction of their total body mass by slightly arching their dorsum, which serves to save energy.

While the bony nasal passages and openings are paired in all cetaceans, the external muscular nasal opening retains its original paired character only in baleen whales. In the toothed whales the narrow leathery partition between the two nasal openings is reduced, giving rise to an unpaired "blowhole." In addition, the soft nasal passages of the toothed whales are widened, still before they enter the bony skull, to form sizable airbags inside a connective vault of the anterior part of their head, the so-called melon. In this way the hydrodynamic head of most toothed whales obtains a drop-like contour, which, beyond doubt, serves to further reduce the resistance in swimming. A unique position of the blowhole is found in the sperm whales. Their unpaired nasal opening is situated asymmetrically in front on the left side of their giant forehead containing the spermaceti organ (cf. Sect. 7.4). However, even this place is the highest point of the whole body, oriented dorsally (Figs. 62, 63).

8.2 Position of the Nostrils and the Mode of Surfacing in Ichthyosaurs

The body of most dolphins and whales is characterized by a pronounced streamlined shape that minimizes water resistance when swimming and enables them to glide through it fluently. An almost identical streamlined shape of body has evolved as a result of the same extreme swimming adaptations independently in the representatives of other vertebrate classes as well (fast-swimming sharks, tuna, swordfish, sailfish, and ichthyosaurs). The contour lines of the body shape of these big, highly effective swimmers show a surprising similarity to the so-called laminar profiles, technically developed so that water or air can flow along them in a laminar way, i.e., without producing disturbing turbulence and thus free of resistance (Gray 1936; Parry 1949; Kramer 1969; Hertel 1963; Klima 1992). Especially the external similarity of the recent dolphins and the fossil ichthyosaurs has repeatedly been pointed out, and the many conformities of the two groups provide a typical example of convergent adaptation to aquatic life habits (Fraas 1905; Abel 1927; Wiman 1943). Often the ichthyosaurs have been denoted as the dolphins of the Mesozoic oceans.

The way in which the cetacean nose was shifted, in the course of evolution, from the tip of the snout up to the top of the head is among the most perfect adaptations to aquatic life. Therefore, it is surprising at first that the ichthyosaurs, in which some adaptations to aquatic life developed convergently with the cetaceans, and which possessed equally perfect adaptations in many a sense, would be thoroughly different from the cetaceans in this point. Their nostrils were situated laterally on the skull, directly in front of the eyes (Fig. 64). However, accepting that the ichthyosaurs, so successful at their time, must have been optimally constructed reptiles even in this

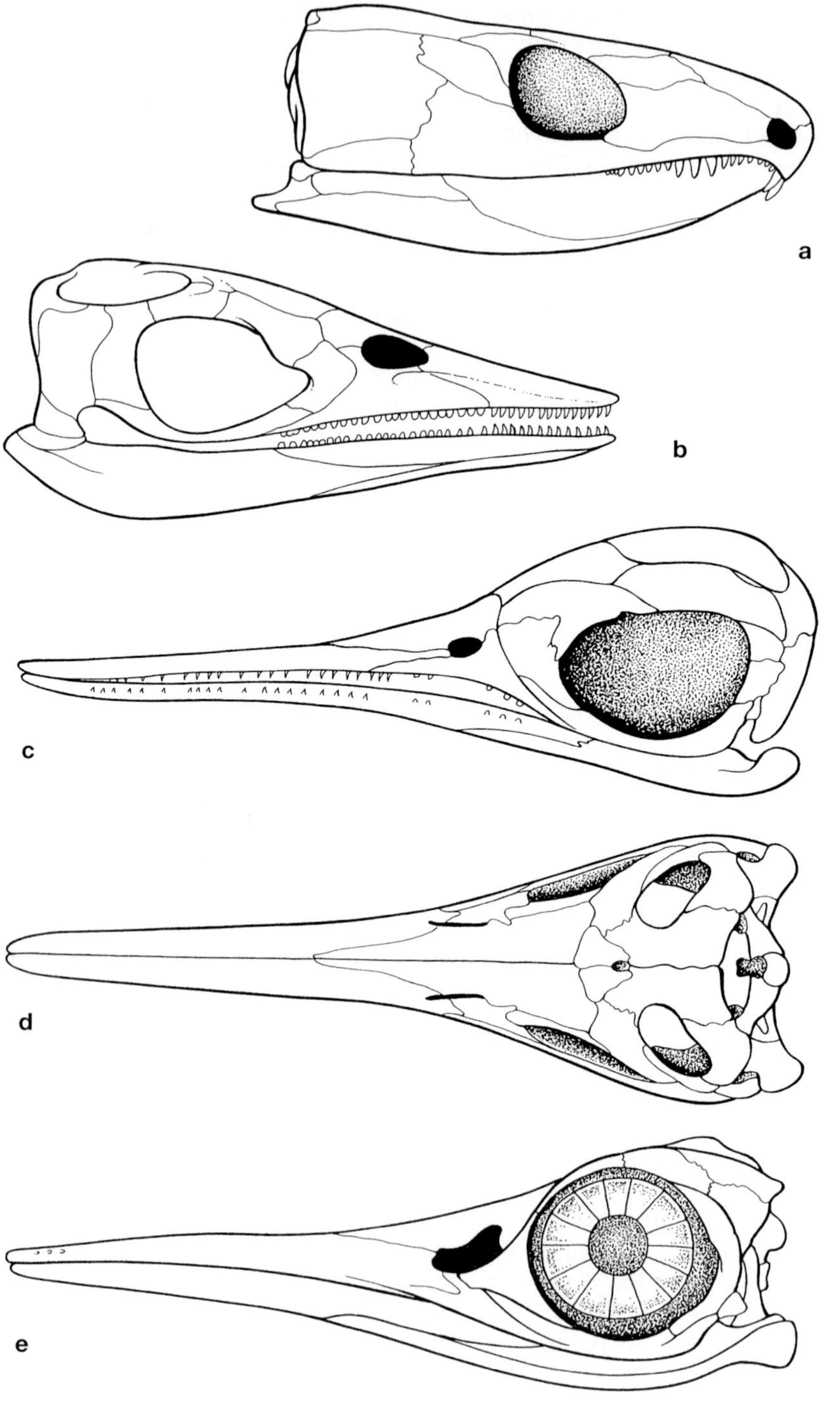
a
b
c
d
e

point, the question arises of what advantage this anatomical construction was. It must be directly connected with the way in which the ichthyosaurs surfaced.

The lateral position of the nostril in the ichthyosaurs is in no way a lower evolutional level than the cetacean nostrils posited on the top of the head. This lateral position is by no means a "not-yet-quite-perfect" position. It is perfect and was final to some extent. Considering that if took the cetaceans about 40 to 50 million years to reach their present evolutional level and that the ichthyosaurs lived for a total period of about 120 to 150 million years, that is, about three times as long, one must admit that they had more than enough time to develop equally perfect adaptations. In this sense, the lateral nostrils lying close in front of the eyes are certainly such perfect adaptations. Their perfection should only be sought for in another connection than cetaceans.

The connection has something to do with the basically different ways in which mammals and reptiles move. In cetaceans the nostrils have been translocated to the vertex of the head as a result of the typical mammalian movement in the dorsoventral direction. Their trunk undulates in verticals; consequently, their horizontally posited tail fin also strikes in verticals. The position of their nose in the highest point of their head enables them to emerge, for breathing out and in, with only a minute part of their whole body. In this, the continuous locomotion is not interrupted, which is energetically advantageous.

In ichthyosaurs the nasal openings have been shifted from the tip of the snout backwards laterally, up to close in front of the eyes. Even this process can be demonstrated palaeontologically (McGowan 1972, 1973a,b; Taylor 1978a,b; Carroll 1988; Wade 1990). The translocation may have developed as a result of the typical reptilian movement in the bilateral direction. The trunk of the ichthyosaurians undulated in horizontals and, consequently, their vertically posited tail fin also strokes in horizontals. To maintain this natural motion even during surfacing, the animals had to turn their longitudinal axis and surface swimming on one side (Fig. 65). In this position, either the left or the right nostril was in the highest place on the head. In this position, too, they were able to emerge with only a small fraction of their whole body, without slowing down or interrupting the fluent locomotion and without any undue increase in energy consumption (Klima 1992, 1993). An analogical example of position changed in surfacing is provided by river dolphins of the genus *Platanista*, which swim in an lateral position and turn into the ventral one when surfacing (Pilleri 1970; Pilleri et al. 1970, 1971).

The ichthyosaurian mode of surfacing in the lateral position is the only thinkable one in which the natural locomotion can be correlated with the position of the nostril

Fig. 64a-e. The nostrils (*black*) of ichthyosaurs are translocated in the course of evolution from the tip of snout posteriorly to the base of the rostrum. They lie close in front of the orbitae and are oriented towards the lateral side . **a** *Captorhinus*, a reptilian ancestor from the Lower Permian; **b** *Grippia*, an ancient form of ichthyosaurs from the Lower Triassic; **c** *Mixosaurus* from the Middle Triassic; **d** *Ichthyosaurus* from the Lower Jurassic; **e** *Ophthalmosaurus*, an advanced form of ichthyosaurs from the Upper Jurassic (From various sources, redrawn by H. Schneeberger)

Fig. 65. Swimming motions and mode of surfacing in the ichthyosaur *Stenopterygius quadriscissus*, according to the newly formulated hypothesis. Swimming in the ventral position when under water, the animals switch to the lateral position when surfacing. Only in this position was one of the laterally situated nasal apertures in the highest place of the body. The body emerged with only a fraction of its total mass and the locomotion proceeded fluently and saved energy. Even leaping in the air was possible in this position (Original painting by M. Klima)

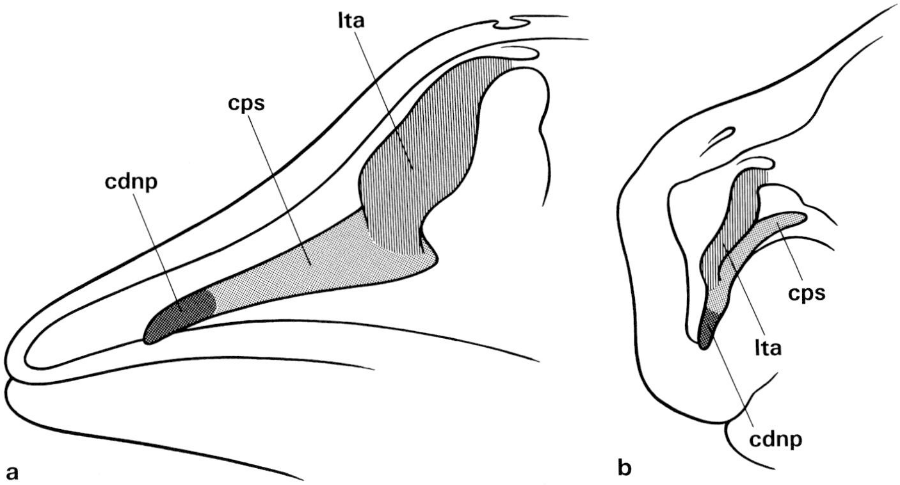

Fig. 66. Comparison of the embryonic nasal structures of dolphins (**a**) with those of sirenians in skull of dugong, *Halicore dugong* (**b**) (The skull of dugong altered from Matthes 1921a)

and the minimum energy expenditure. This mode of surfacing also offers a plausible explanation why the nostril of the ichthyosaurs shifted, during evolution, from the snout tip posteriorly and laterally, close in front of the eyes and not perhaps towards the upper dorsal region at the top of the head as in cetaceans (Klima 1992, 1993).

8.3 Position of the Nostrils and the Mode of Surfacing in Sirenians

The basic correlation between the position of nostrils on the top of the head and the dorsoventral movements of the mammalian body is evidenced, among others, by the sirenians. Besides the cetaceans, they are the only mammalian group that is completely adapted to aquatic life habits and can no longer abandon their aquatic habitat. Sirenians also move by means of trunk and tail movements and their tail fin is positioned horizontally as in cetaceans. Even they have lost their posterior limbs completely. However, sirenians are slow, even lazy swimmers with a plump body. Their head is short and rounded and lacks the long and pointed rostrum typical of whales and dolphins. Yet even their nostrils have moved from the tip of the snout far posteriorly and dorsally. The way sirenians surface is the same as that of cetaceans. The elevated place with the nose emerges as the only part of their body from the water surface when the animals breathe out or in.

At the same time, the embryonic skull of the sirenians shows that, basically, the shift of the nostrils can be due to other morphogenetical processes than in cetaceans. The nostrils are shifted from the snout tip towards upper posterior in front of the flattened forehead but not up to the vertex of the skull (Fig. 66). At the same time, the upper edge of the embryonic septum nasi forms a protruding tubercle accompanied by the lamina transversalis anterior transformed in the same sense. The nostrils open towards the exterior in this highest place. The short rostrum is bent towards the anterior ventral side , also accompanied by the lamina transversalis anterior. The foremost tip of the rostrum is joined by the cartilago ductus nasopalatini. The cartilago paraseptalis behaves quite differently from that of cetaceans. It forms no continuous bar with the lamina transversalis anterior and the cartilago ductus nasopalatini but separates from the lamina transversalis anterior like a branch pointing downwards posteriorly (Freund 1908; Matthes 1912,1921a; De Beer 1937). In this way the cartilago paraseptalis retains largely the original position as in the skull of most land mammals. The decisive structures that account for the translocation of the nostrils in the sirenian skull include solely the septum nasi and the lamina transversalis anterior.

9 The Systematics of Cetacea According to the Morphogeny of Their Nasal Skull

9.1 On the Phylogenetic Origin of Cetacea

The phylogenetic origin of Cetacea is derived from Mesonychidae, a group of Condylarthra from the Palaeocene, from which the first ancient cetaceans (Archaeoceti) arose as a side line of hoofed mammals about 50 million years ago (Van Valen 1968; Szalay 1969; Thenius 1969, 1987; Gaskin 1976, 1982; Barnes and Mitchell 1978; Fordyce 1980, 1988; Barnes et al. 1985). The close relationship of cetaceans to ungulates, especially to artiodactyls, had already been suspected by Flower (1883) and was later confirmed by serological (Boyden and Gemeroy 1950) and molecular biological (Milinkovitch 1992a,b) findings.

Until quite recently, the question of whether all Cetacea actually derive from a single group of terrestrial ancestors has been much disputed. Some authors have considered all conformities in the physical structure of toothed and baleen whales to be mere convergent adaptations to their common holaquatic life habits. This would mean that the Cetacea present no natural unit and that their origin is poly- or diphyletic (Slijper 1962; Yablokov 1964; Rice 1977, 1984; Kuzmin 1976; etc.). This opinion is based, above all, on several weighty differences in the structure of their skulls and dentition as well as on the presence of an echolocation system in Odontoceti or its absence in Mysticeti. Despite these and other differences between the two groups, the hypothesis of the diphyletic origin of cetaceans was not confirmed even by more detailed investigations. On the contrary, there is every reason to believe that Cetaceans originated monophyletically. According to the hitherto paleontological findings, all cetaceans can be unequivocally derived from the archaeocetan Protocetidae, which again can be traced back to a single group of terrestrial ancestral forms, the Mesonychidae (Rothausen 1968; Gaskin 1976, 1982; Fordyce 1980, 1988; Barnes 1984; Barnes et al. 1985). Above all, the characteristic bony otic capsule, the tympanoperioticum, showing the same principal structure in the Archaeoceti, Odontoceti and Mysticeti, could hardly have independently developed three times in the same form, and thus it is the most important key argument for the monophyletic origin of cetaceans (Van Valen 1968; Fleischer 1976, 1978; Oelschläger 1978, 1986a,b, 1990; Pilleri et al. 1987).

Likewise, the cytogenetical investigations made in the 1970s support the hypothesis of the monophyletic origin of cetacean (Árnason 1969, 1970, 1974; Kulu et al. 1971; Árnason and Benirschke 1973; Árnason et al. 1977). They have indicated such extensive identity in Odontoceti and Mysticeti as regards the morphology of chromosomes and the characteristic distribution of their heterochromatin C that a separate origin of these common traits appears to be virtually impossible. It is interesting to note, however, that sperm whales (Physeteridae) show a slight deviation from this scheme. In cytological respect, they show greater similarity to the Mysticeti than to the other Odontoceti, which may indicate their very early phylogenetical separation.

9.2 The Present Results and Their Significance for the Systematics of Cetacea

As indicated by the results described in the present paper, the nasal skull of cetaceans is characterized by profound changes against the original nasal skull of land mammals. As a result of the adaptation to life in water, many elements of the original nasal skull are reduced, above all, the inner nasal structures (conchae, etc.). The whole nasal capsule is heavily simplified (due to the loss of the olfactory function; Oelschläger and Buhl 1985a,b) and, in connection with the translocation of the nasal apertures to the vertex of the head, the nasal capsule turns from horizontal to vertical. At the same time, the original nasal floor is turned, translocated anteriorly and heavily increased in size. Together with the septum nasi that protrudes far anteriorly, the nasal floor becomes a decisive factor in the formation of the more or less long cetacean rostrum. All these transformation processes are distinctly visible as part of skull morphogeny in the course of embryonic development (De Burlet 1916; Honigmann 1917; Eales 1950; Klima and Van Bree 1985, 1989; Klima et al. 1986; Klima 1987a, 1992, 1993). These investigations have revealed that already the earliest anlage of the cetacean nasal skull distinctly differs from that of land mammals in the sense mentioned above. At this early stage the morphology of the nose is identical in all cetacean species examined. Some group-specific modifications and differences do not occur until a late stage when the transformation processes are completed. Accordingly, it is possible to distinguish three basic types of the nasal skull (Fig. 67):

1. *Baleen whales.* They show only a moderate erection of the nasal capsule and a weak reduction of the inner nasal structures, but, on the contrary, a heavy simplification of the nasal floor and a septum nasi protruding very far towards anteriorly. There is a connection with the neurocranium in the form of a massive paired orbitonasal commissure.
2. *Sperm whales.* Here the basal part of the nasal capsule is erect and dorsally tapering to form a narrow cartilaginous bar pointing anteriorly. It is of decisive importance for the longitudinal and altitudinal growth of the big soft forehead and persists even in adults as a long cartilaginous strut that passes diagonally along the whole forehead. The inner structures of the nose are heavily reduced, however not to the extent as in other toothed whales. The nasal floor is heavily simplified. The septum nasi is protracted anteriorly into a long but weak process. The orbitonasal commissures are doubled. The whole anlage shows a conspicuous asymmetry.
3. *Toothed whales* (except sperm whales). Here a steep erection of the nasal capsule and a heavy reduction of the inner nasal structures are observed; on the contrary, the floor of the nasal capsule is only partly reduced, i.e., it still possesses several original elements of the nasal capsule of land mammals. The septum nasi remains relatively short. The connection with the neurocranium through an orbitonasal commissure is only poorly developed or wholly reduced.

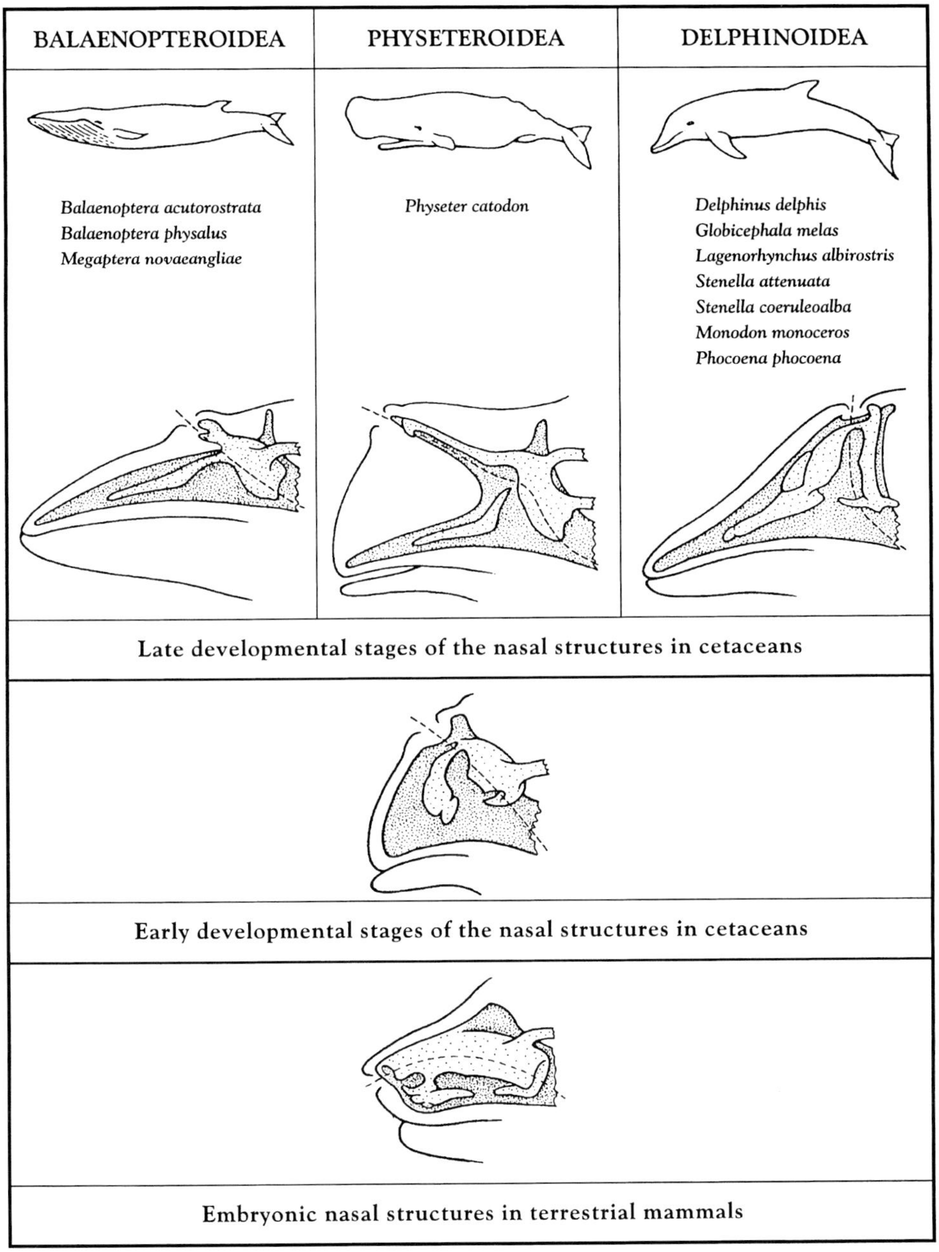

Fig. 67. Division of Cetacea into three main groups according to the development of the cartilaginous structures of the embryonic nasal skull (from Klima 1995)

The fact that in all cetaceans the nasal skull is heavily transformed already at the early fetal stages again emphasizes the common origin of Cetacea. All common traits mentioned above indicate that here we have a single order. The modified nasal skull of the late stage shows, on the one hand, many identical traits but, on the other, several characteristic differences according to which three basic forms can be distinguished, suggesting the division of Cetacea into three major groups. These groups may occupy the rank of superfamilies: Balaenopteroidea, Physeteroidea and Delphinoidea. According to these investigations there is no unequivocal boundary between the baleen and the toothed whales. On the contrary, the embryonic nasal skull of the sperm whale shows, in several characteristics, greater similarity with that of the baleen whales than with that of the toothed ones (Klima and Oelschläger 1994; Klima 1995).

9.3 Other Recent Criteria for the Systematics of Cetacea

Anatomical investigations of the soft structures of the nasal region, especially of the melon, the accessory airbags and the musculature around the nasal apertures, tend to indicate that the sperm whale differs from all other toothed whales and that the differences between the baleen and the toothed whales are much smaller than expected (Mead 1975; Heyning 1989; Heyning and Mead 1990). At the same time, it has appeared as a surprise that the melon, a fatty body in front of the blowhole, which was thought to occur only in toothed whales, is also present in the baleen whales, even if less distinct. This observation tends to falsify the assumptive function of the melon as an "acoustic lens" in emitting ultrasound signals (Norris 1964; Norris et al. 1971; Evans 1973; etc.). The fat-containing tissue may originally have served other purposes, for example to seal the air passages after inspiration, and it has only secondarily, if at all, become a component of the sonar system of toothed whales (Heyning and Mead 1990). However, this observation also increases the importance of those hypotheses which consider not the soft structures but the cartilaginous and osseous tissues of the skull to be the sound-transmitting elements (Purves 1966; Pilleri et al. 1983; Purves and Pilleri 1983; Pilleri 1990). Independent of these considerations, the presence of a miniature melon in baleen whales distinctly indicates their much closer relationship with the toothed ones than believed so far.

Also, the embryological investigations on the pectoral girdle and sternum of cetaceans (Klima 1978, 1990b) lead to the division into the three groups mentioned in Sect. 9.2. Thus, the clavicle cannot be traced at all during the embryogeny of baleen whales; in the sperm whale, it occurs in just one stage as a minute rudiment; in all other toothed whales, on the contrary, its anlage is distinct and regularly developed, and the clavicle can even persist in adult toothed whales (Behrmann 1982; Klima 1990b). It should be noted, however, that no embryonic material of river dolphins (Platanistidae) and beaked whales (Ziphiidae) has been available for examining their nasal skulls or pectoral girdles, which could incidentally still enlarge this division.

The molecular biological investigations made by Irwin et al. (1991), Novacek (1993a,b) and above all, Milinkovitch et al. (1993) have not only confirmed, in many respects the morphologic-systematic connections mentioned above, but they have led to quite unconventional conclusions in several points. The DNA sequences of two ribosomal gene segments from the mitochondria have been examined in 16 species of

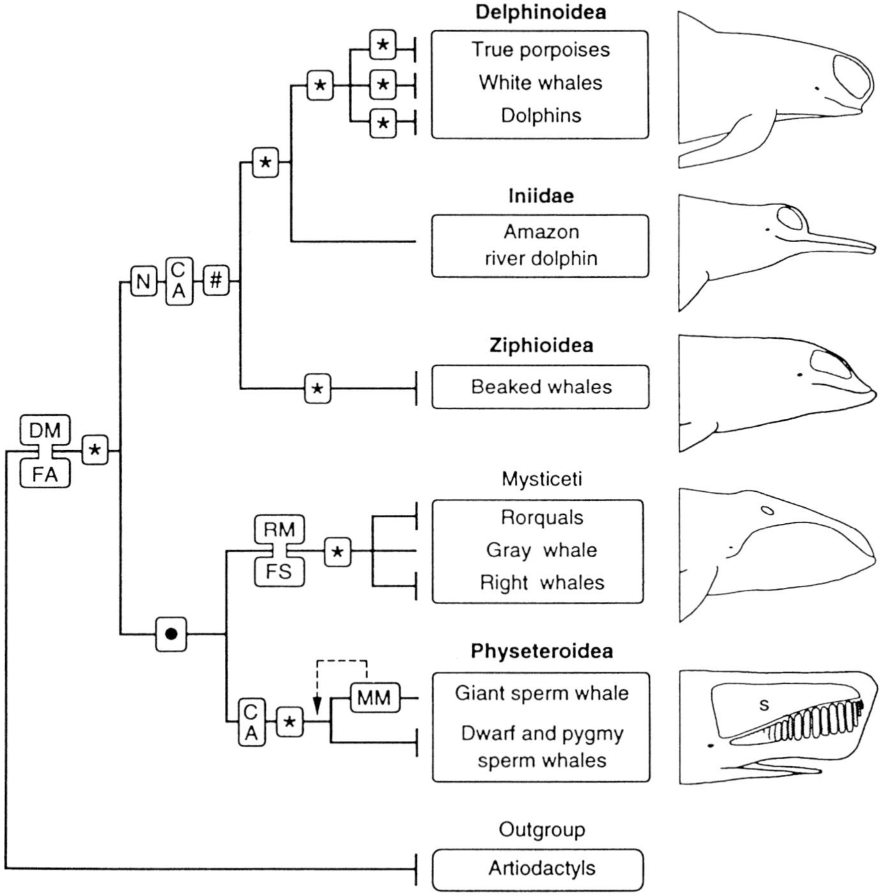

Fig. 68. Cladogram of all extant major groups of cetaceans according to the molecular data (from Milinkovitch 1995)

whales and dolphins, and the results obtained have served to deduce a revised phylogeny of Cetacea (Milinkovitch et al. 1993). First of all, the results confirm a close relationship between Artiodactyla and Cetacea and indicate that all recent superfamilies and families of Cetacea actually are of monophyletic origin. Mutual comparison of the families have revealed that according to these criteria the sperm whales may be more closely related with the baleen whales than with the remaining toothed whales. Consequently, the Odontoceti (in case they should still include the Physeteridae) would be of "paraphyletic" origin, i.e., an artificial group comprising forms that are not mutually related and that have undergone parallel evolution. This would mean that the "toothed whales" are in fact non-existent as a natural phylogenetical group and that its systematic classification requires reconsidering. This would also affect the group of baleen whales, since the traditional division of the order Cetacea into two suborders, Odontoceti and Mysticeti, would become obsolete. The authors have also calculated that the common ancestral form of the baleen and the sperm whales would have lived about 10 to 15 million years ago (but cf. Fordyce 1989, 1992). Besides, this ancestor would have to have possessed a sonar system that was only preserved in the sperm whale but was reduced in the baleen whales. On the whole, the molecular biological investigations of ribosomal gene segments lead to a division of Cetacea into four major groups, however without any representative of river dolphins having been investigated. The molecular biological investigations on the myoglobin in 10 species of Cetacea (Milinkovitch et al. 1993) have led to much the same division. All these results are in full accordance with our observations on the early development of the nasal skull as reported in the present paper. The most recent molecular biological investigations (Milinkovitch 1995) make it possible to deduce a cladogram of all extant major groups of cetaceans, also including the representative of river dolphins, and dividing them into five main groups (Fig. 68). However, the actual significance of the molecular biological findings for systematics must be decided upon through many-sided evolutional-biological analysis in the manner of Henning (1966), and as attempted by Novacek (1993b) for the higher taxa of mammals.

10 Summary

The adaptation of cetaceans to aquatic life habits is reflected, in their nasal region, in three marked changes from the original relations found in land mammals. These changes include (1) the loss of the sense of smell, (2) translocation of the nostrils from the tip of the rostrum to the vertex of the head, and (3) elongation of the anterior head to form a rostrum protruding far towards anterior. The morphogenetic processes taking place during embryogenesis of the nasal skull play a decisive part in the development of all these changes. The lateral parts of the embryonic nasal capsule, encompassing the nasal passages, change their position from horizontal to vertical. At the same time, the structures of the original nasal floor (the solum nasi) are shifted in front of the nasal passages towards the rostrum. The structures of the original nasal roof (the tectum nasi) and of the nasal side wall (the paries nasi) are translocated behind the nasal passages towards the neurocranium. The medial nasal septum (the septum nasi) mostly loses its connection to the nasal passages and is produced into a point protruding far towards anterior. The transformed embryonic nasal skull of the Cetacea can be divided into three sections:

1. The median structures. These include the cartilaginous structures, viz., the rostrum nasi, the septum interorbitale and the spina mesethmoidalis, which are accompanied by the dermal bones, the vomer and the praemaxillare. In adult cetaceans the rostrum nasi is mostly preserved as a robust cartilage of the skull, which may possibly serve as a sound transmitting structure of the sonar system, or it may be responsible for the sensing of water streams and vibrations.

2. The posterior side wall structures. These include the following cartilaginous structures that are mostly heavily reduced or mutually fused: the cupula nasi anterior, the tectum nasi, the lamina cribrosa, the paries nasi, the commissura orbitonasalis, the cupula nasi posterior, the processus paraseptalis posterior, the crista semicircularis, the frontoturbinale, the ethmoturbinale I and the maxilloturbinale. The cartilaginous structures are largely accompanied by the dermal bone, the maxillare. Of these embryonic elements, very little is preserved in adult cetaceans. The cartilages of the cupula nasi anterior form the variable skeleton around the nostrils. In *Physeter* the tectum nasi forms a very long cartilaginous bar that passes through the whole giant anterior head of the sperm whale as a structure accompanying the left nasal passage.

3. The anterior side wall structures. These include the cartilaginous structures, viz., the cartilago ductus nasopalatini, the cartilago paraseptalis, the processus lateralis ventralis and the lamina transversalis anterior, accompanied by the dermal bones, the praemaxillare and the vomer. These structures participate in the formation of the robust rostrum of the cetacean skull, and they are partly preserved even in adults in the form of the isolated ossa pararostralia (the Meckelian ossicles).

The comparison of morphogeny of the nasal skull has also made it possible to draw certain conclusions on the phylogeny and systematics of Cetacea. Already the earliest embryonic stages permit us to discern weighty transformations of the original nasal skull of land mammals. These transformations are common to all embryos examined. This fact indicates a common origin of all Cetacea, which thus form a single monophyletic order. However, later embryonic stages show some different modifications of the nasal capsule according to which at least three major groups can be distinguished within the order Cetacea, probably ranking as superfamilies: Balaenopteroidea, Physeteroidea and Delphinoidea. Our observations, being in full accordance with other morphological, and embryological, as well as molecular biological results, suggest that the division of the order Cetacea into two suborders, Mysticeti and Odontoceti, is no longer tenable.

Acknowledgements. I am deeply indebted to all individuals and institutions for access to the material investigated, especially to Dr. G.C. Bangma, Utrecht; G. Behrmann, Bremerhaven; Dr. E.C. Boterenbrood, Utrecht; P. Deimer, Dipl. Biol., Hamburg; Dr. A. Jonsgard, Oslo; Prof. Dr. T. Kamiya, Tokyo; Dr. W.M. Marquette, Seattle; Dr. W.P. Perrin, La Jolla; and Dr. D.W. Rice, Seattle.

My sincere thanks go to Dr. R. Obrtel, Brno, for translating most of the manuscript.

I am most grateful to I. Szász and H.L. Schneeberger, both of Frankfurt, for drawing the models and for making some other drawings. It is a pleasure to thank I. Kirschenbauer, Frankfurt, for technical assistance and for the typing of the manuscript. My gratitude is also extended to I. Pflug, Frankfurt, for typing some parts of the manuscript.

Last but not least I would like express my thanks to „Beauty“, our cat, who was my companion during the long days and nights of writing the manuscript.

References

Abel O (1902) Die Ursache der Asymmetrie des Zahnwalschädels. Sitz-Ber Kais Akad Wiss Wien, Math-Naturw Kl 111 (1):510–526

Abel O (1927) Lebensbilder aus der Tierwelt der Vorzeit. - VII + 7145, 551 Abb, Jena (Fischer)

Amundin M, Cranford TW (1990) Forehead anatomy of Phocoena phocoena and Cephalorhynchus commersonii: 3-dimensional computer generated reconstructions with emphasis on the nasal diverticula. In: Sensory abilities of Cetacea, Thomas J and Kastelein R, (eds) Plenum, New York, pp 1–18

Árnason Ú (1969) The karyotypes of the fin whale. Hereditas 62:273–284

Árnason Ú (1970) Karyotypes of a male sperm whale (Physeter catodon L.) and a female sei whale (Balaenoptera borealis Less.). Hereditas 64:237–242

Árnason Ú (1974) Comparative chromosome studies in Cetacea. Hereditas 77:1–36

Árnason Ú, Benirschke K (1973) Karyotypes and idiograms of sperm and pygmy whales. Hereditas 75:67–74

Árnason Ú, Benirschke K, Mead JG, Nichols WW (1977) Banded karyotpyes of three whales: Mesoplodon europaeus, M. carlhubbsi and Balaenoptera acutorostrata. Hereditas 87:189–200

Ärnbäck Christie-Linde A (1914) On the cartilago palatina and the organ of Jacobson in some mammals. Morph J 48:343–364

Arvy L, Pilleri G (1983) Le cachalot: Physeter macrocephalus (Linné 1758). In: Quatre cents ans d'observations sur l'hyperoodon, le cachalot et l'orque. Investig. Cetacea 15 (Suppl):81–191

Barnes LG (1984) Search for the first whale: retracing the ancestry of cetaceans. Oceans 17:20–23

Barnes LG, Domning LG, Ray CE (1985) Status of studies on fossil marine mammals. Mar Mam Sci 1:15–53

Barnes LG, Mitchell ED (1978) Cetacea. In: Maglio VJ and Cooke HBS (eds): Evolution of African mammals. Harvard University Press, Cambridge, MA

Beddard FE (1915) Contributions to the knowledge of the anatomy of the sperm whale (Physeter macrocephalus) based upon the examination of a young foetus. Ann Durban Mus 1:107–124

Beddard FE (1919) Further contributions to the anatomy of the sperm whale (Physeter macrocephalus) based upon an examination of two additional foetus. Ann Durban Mus 2:129–148

Behrmann G (1982) Schlüsselbeine beim Schwertwal, Grampus orca, Schaltfehler oder nicht? Der Präparator 28:201–204

Behrmann G (1985) Kurzer Vorbericht über den in Bremerhaven angelandeten Pottwal, Physeter macrocephalus Linné 1758. Der Präparator 31:59–62

Behrmann G (1989) The cartilaginous rostrum and the associated rostral sense-organ of toothed whales (Odontoceti). Lutra 32:152–163

Behrmann G, Klima M (1985) Knorpelstrukturen im Vorderkopf des Pottwals Physeter macrocephalus. Z f Säugetierkunde 50:347–356

Bersch W, Reinbach W (1970) Das Primordialcranium eines menschlichen Embryo von 52 mm Sch.-St.-Länge. Zur Morphologie des Craniums älterer menschlicher Feten II. Z Anat Entwickl Gesch 132:240–259

Berzin AA (1971) The sperm whale. Israel program for scientific translations, Jerusalem 1972 (transl by Hoz E and Blake Z)

Boenninghaus G (1903) Der Rachen von Phocaena communis. Less Zool J Anat 17:1–98

Boyden A, Gemeroy D (1950) The relative position of the Cetacea among the orders of Mammalia as indicated by precipitin tests. Zoologica 35:145–151
Broom R (1896) On the comparative anatomy of Jacobsons organ in marsupials. Proc Linn Soc NSW 21:591–623
Broom R (1909) Observation on the development of the marsupial skull. Proc Linn Soc NSW 34:195–214
Buhl EH, Oelschläger HA (1986) Ontogenetic development of the nervus terminalis in toothed whales. Evidence for its non-olfactory nature. Anat Embryol 173:285–294
Buhl EH, Oelschläger HA (1988) Morphogenesis of the brain in the harbour porpoise. J Comp Neurol 227:190–225
Carroll RL (1988) Vertebrate paleontology and evolution. XIV 6985, New York (Freeman)
Cave AJE (1987) The Meckelian ossicle of odontocete cranium. Investigat Cetacea Bern 20:178–191
Cave AJE (1988) The os maxilloturbinare in the cetacea. Investigat Cetacea Bern 21:41–50
Clarke MR (1970) Function of the spermaceti organ of the sperm whale. Nature 228:873–874
Clarke MR (1978a) Structure and proportions of the spermaceti organ in the sperm whale. J Mar Biol Ass UK 58:1–17
Clarke MR (1978b) Buoyancy control as a function of the spermaceti organ in the sperm whale. J Mar Biol Ass UK 58:27–71
Clarke MR (1979a) The head of the sperm whale. Scientific American 240:106–117
Clarke MR (1979b): Der Kopf des Pottwals. Spektrum der Wissenschaft 21–28 (übersetzt von Horn I)
Cords E (1915) Über das Primordialcranium von Perameles spec (?) unter Berücksichtigung der Deckknochen. Anat Hefte 1 Abt 52:1–83
Cousteau J (1988) Whales. Abrams, New York
Cox V (1990) Wale und Delphine. Karl Müller Verlag, Erlangen
Cranford TW (1988) The anatomy of acoustic structures in the Spinner dolphin forehead as shown by X-ray computed tomography and computer graphics. In: Animal Sonar, Processes and performance, (eds) Nachtigall PE and Moore PWB, Plenum, New York, pp 67–77
Cranford TW, Amundin M, Norris KS (1996) Functional morphology and homology in the odontocete nasal complex: implications for sound generation. J Morph 228:223–285
De Beer GR (1937) The development of the vertebrate skull. Clarendon Press, Oxford
De Beer GR, Fell WA (1936) The development of the monotremata. - Part III. The development of the skull of Ornithorynchus. Trans Zool Soc Lond 23:1–42
De Beer GR, Woodger JH (1930) The early development of the skull of the rabbit. Phil Trans Roy Soc London 218:373–414
De Burlet HM (1913a) Zur Entwicklungsgeschichte des Walschädels. I. Über das Primordialcranium eines Embryo von Phocaena communis. Morph Jb 45:523–556
De Burlet HM (1913b) Zur Entwicklungsgeschichte des Walschädels. II. Das Primordialcranium eines Embryo von Phocaena communis von 92 mm. Morph Jb 47:645–676
De Burlet HM (1914a) Zur Entwicklungsgeschichte des Walschädels. III. Das Primordialcranium eines Embryo von Balaenoptera rostrata (105 mm). Morph Jb 49:119–178
De Burlet HM (1914b) Zur Entwicklungsgeschichte des Walschädels. IV. Über das Primordialcranium eines Embryo von Lagenorhynchus albirostris. Morph Jb 49:393–406
De Burlet HM (1916) Zur Entwicklungsgeschichte des Walschädels. V. Zusammenfassung des über den Knorpelschädel der Wale Mitgeteilten. Morph Jb 50:1–18
Decker F (1883) Über den Primordialschädel einiger Säugetiere. Zeitschr f wissenschaftl Zoologie 38
Denison W, Terry RJ (1921) Washingt Univ Stud 8:161
Dierbach AR (1985) Zur Morphogenese des Craniums von Cavia porcellus L. Teil I. Einführung, Systematik und beschreibender Teil. Morph Jb 131:441–476 - Teil II. Vergleichender Teil und Schrifttum. Morph Jb 131:617–642
Eales NB (1950) The skull of the fetal narwhal Monodon monoceros. L Phil Trans Roy Soc London 235B 621:1–33
Eloff FC (1948) The early development of the skull of Otomys tropicalis. Ann Transv Mus 21:103–152
Eloff FC (1950) On the nasal region of the chondrocranium of the cape hare, Lepus capensis. Ann Transv Mus 21:222–233
Eloff FC (1951a) Observation on the Chondrocranium of Rhabdomys pumilio. Ann Transv Mus 21:369–379

Eloff FC (1951b) On the organ of Jacobson and the nasal-floor cartilagines in the chondrocranium of Galago senegalensis. Proc Zool Soc London 1951:651–655
Eschricht DF (1849) Undersögelser over Hvaldyrene. Copenhagen. Zool Anat-Phys Untersuchungen über die nordischen Wahlthiere. Leipzig
Eschricht DF, Reinhardt J (1861) Om Nordhvalen Balaena mysticetus L. Kongelike Danske Videns Kabernes Selskab Schrifter, 5th Rankke, Naturvidens-Kabelig og Mathematisk Afdeling, 5th Bind 1–160
Esdaile PC (1916) On the structure and development of the skull and laryngeal cartilages in Perameles, with notes on the cranial nerves. Phil Trans Roy Soc London 207:439–479
Evans WE (1973) Echolocation by marine delphinids and one species of freshwater dolphin. J Acoust Soc Amer 54:191–199
Fawcett E (1911) The development of the human maxilla, vomer, and paraseptal cartilages. J Anat Physiol London 45:378–405
Fawcett E (1917) The primordial cranium of Microtus amphibius (water rat), as determined by sections and a model of the 25 mm stage. With comparative remarks. J Anat 51:309–359
Fawcett E (1918a) The primordial cranium of Erinaceus europaeus. J Anat 52:211–250
Fawcett E (1918b) The primordial cranium of Poecilophoca weddelli (Wedell's seal), at the 27 mm C.R. length. J Anat 52:412–441
Fawcett E (1919) The primordial cranium of Miniopterus schreibersi of the 17 millimeter total length stage. J Anat 53:315–348
Fawcett E (1923) The primordial cranium of Xerus (spiny squirrel) at the 17 and 19 millimeter stages. J Anat 57:221–237
Felts W (1966) Some functional and structural characteristics of cetacean flippers and flukes. In: Norris KS (eds): Whales, dolphins and porpoises. Univ Calif Press, Berkeley
Felts WJL, Spurrel FA (1965) Structural orientation and density in cetacean humeri. Amer J Anat 116:171–203
Fischer E (1901) Das Primordialcranium von Talpa europaea. Anat Hefte 17:467–548
Fischer E (1903) Zur Entwicklungsgeschichte des Affenschädels. Z Morph Anthrop 5:383–414
Fischer E (1905) On the primordial cranium of Tarsius spectrum. Proc Kon Ned Akad Wet 8:397–400
Fleischer G (1976) Hearing in extinct cetaceans as determined by cochlear structure. J Paleontol 50:133–152
Fleischer G (1977) Über die Erforschung der Schall-Orientierung bei Delphinen und Walen. Jh Ges Naturkde Württemberg 132:18–34
Fleischer G (1978) Evolutionary principles of the mammalian middle ear. Erg Anat Entwickl Gesch 55:1–70
Fleischer G (1982) Hörmechanismen bei Delphinen und Walen. Dtsch Ges Hals-, Nasen-, Ohrenkde 30:123–130
Flower WH (1883) On whales, present and past and their probable origin. Proc Zool Soc London 466–513
Fordyce RE (1980) Whale evolution and Oligocene southern ocean environments. Palaeogeography, Palaeoclimatology, Palaeoecology 31:319–336
Fordyce RE (1988) Evolution. In: Whales, Dolphins and Porpoises. Harrison R and Bryden MM (eds) Merehurst Press, London, pp 14–23
Fordyce RE (1989) Origins and evolution of Antarctic marine mammals. In: Crame JA (ed) Origins and Evolution of the Antarctic Biota. Geological Society Special Publication 47:269–281
Fordyce RE (1992) Cetacean evolution and Eocene Oligocene environments. In: Protero D and Bergen W (eds) Eocene-Oligocene Climatic and Biotic Evolution. Princeton University Press, Princeton, New Jersey, pp 368–381
Fraas E (1905) Reptilien und Säugetiere in ihren Anpassungserscheinungen an das marine Leben. Jh Ver vaterl Naturk Württ, Stuttgart, 61:347–386
Frangioni G (1970) The chondrocranium of Proteles cristatus and of Genetta genetta. Monitore Zool Ital Suppl 3:173–192
Fraser F, Purves PE (1960) Hearing in Cetaceans. Bull Brit Mus Zoll 7:1–140
Frets GP (1912) Beiträge zur vergleichenden Anatomie und Ontogenie der Nase der Primaten. I. Beobachtungen und Bemerkungen zur Entwicklung der Nase bei einigen catarrhinen Affen, Säugern und dem Menschen. Morph Jb 44:409–463

Frets GP (1913) Beiträge zur vergleichenden Anatomie und Embryologie der Nase der Primaten. II. Die Regio ethmoidalis des Primordialcraniums mit Deckknochen von einigen platyrrhinen Affen. Morph Jb 45:557–726
Frets GP (1914) Beiträge zur vergleichenden Anatomie und Embryologie der Nase der Primaten. III. Die Regio ethmoidalis des Primordialcraniums mit Deckknochen von einigen Catarrhinen, Prosimiae und dem Menschen. Morph Jb 48:239–279
Freund L (1908) Beiträge zur Entwicklungsgeschichte des Schädels von Halicore dugong Erxl. Semon. Zool. Forschungsreisen in Australien. Denkschr med naturw Ges Jena 7:557–626
Freund L (1911) Zur Morphologie des Nasenknorpels. Beitr Anat Phys Pathol Ther d Ohres, d Nase, d Halses 4:414–438
Frick H (1954) Die Entwicklung und Morphologie des Chondrocraniums von Myotis Kaup. (Fischer) Stuttgart
Frick H (1986) Zur Entwicklung des Knorpelschädels der Albinomaus. Nova acta Leopoldina NF 58(262):305–317
Frick H, Heckmann U (1955) Ein Beitrag zur Morphogenese des Kaninchenschädels. Acta Anat 24:268–314
Frick H, Starck D (1963) Vom Reptil- zum Säugetierschädel. Z Säugetierk 28:321–341
Fuchs H (1909) Über die Entwicklung einiger Deckknochen (Vomer, Pterygoid, Maxillare) bei Säugetieren (und ihr Verhältnis zum Knorpelskelette). Verh Anat Ges, Jena 23:85–104
Gaskin DE (1976) The evolution, zoogeography and ecology of cetacea. Oceanogr Mar Biol Ann Rev 14:247–346
Gaskin DE (1982) The ecology of whales and dolphins. Heinemann, London/Exeter
Gaupp E (1893) Beiträge zur Morphologie des Schädels. I. Primordial-Cranium und Kieferbogen von Rana fusca. Morphol Arbeiten Schwalbe G (ed) 2:275–481
Gaupp E (1900) Das Chondrocranium von Lacerta agilis. Ein Beitrag zum Verständnis des Amniotenschädels. Anat Hefte 15:433–595
Gaupp E (1905) Neue Deutungen auf dem Gebiet der Lehre vom Säugetierschädel. Anat Anz 27:273–310
Gaupp E (1908) Zur Entwicklungsgeschichte und vergleichenden Morphologie des Schädels von Echidna aculeata var. typica. Semons zoologische Forschungsreisen in Australien 6:539–788
Gewalt W (1976) Der Weißwal. Die Neue Brehm-Bücherei. A Ziemsen-Verlag, Wittenberg, Lutherstadt
Gray J (1936) Studies in animal locomotion VI. The propulsive powers of the dolphin. J Exp Biol 13:192–199
Gruhl K (1911) Beiträge zur Anatomie und Physiologie der Cetaceennase. Z Naturwiss Jena 47:367–414
Halata Z (1975) The mechanoreceptors of the mammalian skin. Adv Anat Embryol Cell Biol 50:1–77
Halbsguth A (1973) Das Cranium eines Foeten des Flattermaki Cynocephalus volans (Galeopithecus volans) (Mammalia, Dermoptera) von 63 mm SchStlg. Med dissertation Frankfurt/Main
Hauck S (1978) Zur ontogenetischen Entwicklung der Ethmoidal- und Orbitotemporalregion bei Phodopus sungorus. dissertation Fachb Veterinärmed, Universität Gießen
Hauschild R (1937) Rassenunterschiede zwischen negriden und europiden Primordialcranien des 3. Fetalmonats. Z Morph u Anthrop 36:215–279
Henckel KO (1927a) Zur Entwicklungsgeschichte des Halbaffenschädels. Z Morph Anthropol 26:365–383
Henckel KO (1927b) Das Primordialcranium der Halbaffen und die Abstammung der höheren Primaten. Verh Anat Ges 36:108–116
Henckel KO (1928) Das Primordialcranium von Tupaia und der Ursprung der Primaten. Z Anat Entwickl-Gesch 86:204–227
Henckel KO (1929) Die Entwicklung des Schädels von Galeopithecus temmincki Waterh. und ihre Bedeutung für die stammesgeschichtliche und systematische Stellung der Galeopithecidae. Morph Jb (Maurerfestschrift part 1) 62:179–205
Hennig W (1966) Phylogenetic systematics. Univ Illinois Press, Urbana
Hertel H (1963) Struktur, Form, Bewegung; Biologie und Technik. Krausskopf, Mainz
Heyning JE (1989) Comparative facial anatomy of beaked whales (Ziphiidae) and a systematic revision among the families of extant Odontoceti. Contrib in Science, Nat Hist Mus Los Angeles Co 405:1–64

Heyning JE, Mead JG (1990) Evolution of the nasal anatomy of cetaceans. In: Thomas J and Kastelein R (eds) Sensory Abilities of Cetaceans, Plenum Press, New York, pp 67–79

Hirschfelder H (1937) Das Primordialcranium von Manatus latirostris. Z Anat Entw Gesch 106:497–533

Honigmann H (1915) Das Primordialkranium von Megaptera nodosa. Anat Anz 48:113–127

Honigmann H (1917) Bau und Entwicklung des Knorpelschädels vom Buckelwal. Zoologica, Stuttgart 69:1–87

Irwin DM, Kocher TD, Wilson AC (1991) Evolution of the cytochrome b gene of mammals. J Mol Evol 32:128–144

Kadam KM (1972) The development of the skull in the Indian Gerbil, Tatera indica cuviere (Waterhouse) Part I. Morph Jb 118:309–325

Kadam KM (1973a) The development of the skull in the Indian Gerbil, Tatera indica cuviere (Waterhouse) Part II. Morph Jb 119:47–71

Kadam KM (1973b) The development of the skull in the Indian Gerbil, Tatera indica cuviere (Waterhouse) Part III. Morph Jb 119:153–171

Karger S (1988) Zur Morphogenese der Ethmoidal- und Orbitotemporalregion bei Castor fiber. Med dissertation Univ Frankfurt

Kellogg R (1928) The history of whales – their adaptation to life in the water. Q Rev Biol 3:29–76; 174–208

Kellogg R (1936) A review of the Archaeoceti. Carnegie Inst Washington, Publ 482

Kellogg R (1961) Porpoises and sonar. University of Chicago Press, Chicago

Klauer G (1976) Zum Bau und zur Innervation des Nasenspiegels von Tupaia glis (Diard 1820). Diplomarbeit, FB Biologie, Univ Gießen

Klima M (1978) Comparison of early development of sternum and clavicle in striped dolphin and in humpback whale. Sci Rep Whales Res Inst Tokyo 30:253–269

Klima M (1987a) Morphogenesis of the nasal structures of the skull in toothed whales (Odontoceti). In: Kuhn HJ and Zeller U (eds) Morphogenesis of the Mammalian Skull, Mammalia Depicta 13, Beihefte z Z f Säugetierkunde, Paul Parey, Hamburg, pp 105–121

Klima M (1987b) Early development of the shoulder girdle and sternum in marsupials (Mammalia: Metatheria). Adv Anat Embr Cell Biol 109:1–91

Klima M (1990a) Histologische Untersuchungen an Knorpelstrukturen im Vorderkopf des Pottwals Physeter macrocephalus. Gegenbaurs morphol Jahrb 136:1–16

Klima M (1990b) Rudiments of the clavicle in the embryos of whales (Cetacea). Z f Säugetierkunde 55:202–212

Klima M (1992) Schwimmbewegungen und Auftauchmodus bei Walen und bei Ichthyosauriern. Natur u Museum, Frankfurt am Main 122/1:1–17, 3:73–89

Klima M (1993) Über einen möglichen Auftauchmodus bei den Ichthyosauriern. N Jb Paläont, Abh, Stuttgart, 188:389–412

Klima M (1994) Anpassungen an die aquatische Lebensweise. In: Niethammer J and Krapp F (eds) Handbuch der Säugetiere Europas, Bd. 6: Meeressäuger. Aula-Verlag, Wiesbaden, pp 49–79

Klima M (1995) Cetacean phylogeny and systematics based on the morphogenesis of the nasal skull. Aquatic mammals 21(2):79–89

Klima M, Bangma GC (1987) Unpublished drawings of marsupial embryos from the Hill Collection and some problems of marsupial ontogeny. Z f Säugetierkunde 52:201–211

Klima M, Behrmann G (1985) Unbekannte Knorpelstrukturen im Vorderkopf des Pottwals (Physeter macrocephalus). 59. Hauptversammlung Deutsche Gesellschaft für Säugetierkunde (abridged versions) Hamburg, Parey, 31

Klima M, Oelschläger HA (1994) Phylogenie und Systematik der Cetacea. In: Niethammer J and Krapp F (eds) Handbuch der Säugetiere Europas, Bd. 6: Meeressäuger. Aula-Verlag, Wiesbaden, pp 31–48

Klima M, Van Bree PJH (1985) Überzählige Skeletelemente im Nasenschädel von Phocoena phocoena und die Entwicklung der Nasenregion bei den Zahnwalen. Gegenbaurs Morphol Jahrb 131:11–178

Klima M, Van Bree PJH (1990) On the origin of the so-called Meckelian ossicles in the nasal skull of Odontocetes. Gegenbaurs Morphol Jahrb 136:431–434

Klima M, Seel M, Deimer P (1986) Die Entwicklung des hochspezialisierten Nasenschädels von Pottwal (Physeter macrocephalus). Gegenbaurs Morphol Jahrb 132:245-284, 349–374

Köhncke M (1980) Das Chondrocranium von Cryptoprocta ferox - ein Beitrag zur vergleichenden Anatomie der Carnivora. Dissertation des Fachbereichs Biologie, Hamburg
Köhncke M (1985) The Chondrocranium of Cryptoprocta ferox. Adv Anat Embr Cell Biol 95:1–89
Kramer MO (1960a) The dolphin secret. New Scientist London 7:1118–1120
Kramer MO (1960b) Boundary layer stabilization by distributed damping. J Amer Soc Nav Engnrs 72:25–33
Kuhn HJ (1971) Die Entwicklung und Morphologie des Schädels von Tachyglossus aculeatus. Abh senckenb naturforsch Ges 528:1–192
Kükenthal W (1893) Vergleichend-anatomische und entwicklungsgeschichtliche Untersuchungen an Walthieren: Bau und Entwicklung der Cetaceennase. Denkschr Med Naturw Ges Jena 3:322–349
Kulu DD (1972) Evolution and cytogenetics. In: Ridgway SH (ed) Mammals of the Sea: Biology and Medicine. Thomas, Springfield pp 503–527
Kulu DD, Veomett I, Sparkes RS (1971) Cytogenetic comparison of four species of cetaceans. J Mammal 52:828–832
Kummer B, Neiss S (1957) Das Cranium eines 103 mm langen Embryos des südlichen See-Elefanten (Mirounga leonina L.). Morph Jb 98:288–346
Kuzmin AA (1969) Some data on the growth (long measures) of skull of the Sperm Whale in embryogenesis. 4-ed All Union Meeting on Studies of Marine Mammals. Summaries papers, Moscow 119–121
Kuzmin AA (1974) The development of skull bone of Physeter macrocephalus (Cetacea) in the prenatal period. Ist Int Congr on Mammals, Moscow 1:322
Kuzmin AA (1976) Embryogenesis of the osseous skull of the sperm whale (Physeter macrocephalus Linnaeus, 1758). Investig Cetacea 7:187–202
Lacepède BGE (1804) Histoire naturelle des cétacées. Plassan, Paris
Levi G (1900) Beitrag zum Studium der Entwicklung des knorpeligen Primordialcraniums des Menschen. Arch Mikr Anat 55:341–414
Lindahl PE (1948) Über die Entwicklung und Morphologie des Chondrocraniums von Procavia capensis Pall. Acta Zool 29:281–376
Macklin CC (1914) The skull of a human foetus of 40 mm. Amer J Anat 16:317369, 387–426
Macklin CC (1921) The skull of a human foetus of 43 mm greatest length. Contrib Embryol Carneg Inst 10 (48):57–103
Maier W (1981) Nasal structures in old and new world primates. In: Ciochon RL, Chiarelli AB (eds) Evolutionary biology of the new world monkeys and continental drift. Plenum, New York, pp 219–241
Maier W (1983) Morphology of the interorbital region of Saimiri sciureus. Folia Primatol 41:277–303
Maier W (1987) The ontogenetic development of the orbitotemporalregion in the skull of Monodelphis domestica (Didelphidae, Marsupalia) and the problem of Mammalian alisphenoid. Mammalia depicta, Beih Z Säugetierkunde, Kuhn HJ and Zeller U (eds) pp 71–90
Maier W (1991) Aspects of ontogenetic development of nasal and facial skeletons in Primates. In: Pfeifer G (ed) Craniofacial abnormalities and clefts of the lip, alveolus and palate, 4th Hamburg Internat Symposium. Thieme Verlag Stuttgart pp 115–124
Martin AR (1992) Das große Buch der Wale und Delphine. Kremayr u Scherian, Wien und Bertelsmann, Gütersloh
Matthes E (1912) Zur Entwicklung des Kopfskelettes der Sirenen. I. Die Regio ethmoidalis des Primordialkraniums von Manatus latirostris. Jena Z Naturwiss 48:489–514
Matthes E (1921a) Zur Entwicklung des Kopfskelettes der Sirenen. II. Das Primordialcranium von Halicore dugong. Z Anat Entwickl Gesch 60:1–306
Matthes E (1921b) Neuere Arbeiten über das Primordialcranium der Säugetiere. Ergebn Anat Entwickl Gesch 23:669–912
Matthes E (1923) Neuere Arbeiten über das Primordialcranium der Säugetiere. II. Teil. Arbeiten über einzelne Regionen des Craniums. Ergebn Anat Entwickl Gesch 24:117–243
Matthews LH (1938) The sperm whale, Physeter catodon. Discovery Reports 7:93–168
Mayer A (1835) Beiträge zur Anatomie des Delphins. Z Physiol 5:111–133
McGowan C (1972) Evolutionary trends in longipinnate Ichthyosaurs with particular references to the skull and fore fin. Life Sci Contrib Roy Ontario Museum, Toronto 83:22–35

McGovan C (1973a) Differential growth in three Ichthyosaurs: Ichthyosaurus communis, I. breviceps, and Stenopterygius quadriscissus (Reptilia, Ichthyosauria). Life Sci Contrib Roy Ontario Museum, Toronto 93:1–21

McGovan C (1973) The cranial morphology of the lower liassic latipinnate Ichthyosaurs of England. Bull Brit (Nat Hist) Geol, London 24/1:1–109

Mead CS (1909) The chondrocranium of an embryo pig, Sus scrofa. A contribution to morphology of the mammalian skull. Am J Anat 9:167–210

Mead JG (1975) Anatomy of the external nasal passages and facial complex in the Delphinidae (Mammalia: Cetacea). Smithsonian Contrib Zool 207:1–72

Meckel JF (1825) Epistola gratularia ad Blumenbachum. Lipsiae, Halse 1825

Menzel KH (1979) Morphologische Untersuchungen an der vorderen Nasenregion von Solenodon paradoxus (Insectivora). Med dissertation Frankfurt/Main

Milinkovitch MC (1992a) DNA-DNA hybridizations support ungulate ancestry of Cetacea. J Evol Biol 5:149–160

Milinkovitch MC (1992b) Use of molecular techniques in the study of Cetacea phylogeny and discrimination of their populations. In: Symoens JJ (ed) Symposium Whales: Biology-Threats-Conservation, Brussels, 5–7 June 1991, Proceedings. Royal Academy of Overseas Sciences, Brussels pp 45–58

Milinkovitch MC (1995) Molecular phylogeny of cetaceans prompts revision of morphological transformations. Trends Ecol Evol 10, 8:328–334

Milinkovitch MC, Orti G, Meyer A (1993) Revised phylogeny of whales by mitochondrial ribosomal DNA sequences. Nature 361:346–348

Miller GS (1923) The telescoping of the cetacean skull. Smithson Misc Collect 75:1–68

Moore WJ (1981) The mammalian skull. Biological structure and function. Cambridge Univ Press, Cambridge

Muggia G (1931) Der Knorpelschädel eines Pferdeembryos. Z Anat Entwicklungsgesch 95:297–325

Müller D (1981) Beitrag zur Craniogenese von Saguinus tamarin Link, 1795 (Platyrrhini, Primates). Cour Forsch Inst Senckenberg 46:1–100

Müller U (1986) Zur Morphogenese der Ethmoidal- und Orbitotemporalregion bei Wallabia rufogrisea (Marsupalia). Med dissertation Univ Frankfurt

Nishiwaki M, Oshumi S, Maeda Y (1963) Change of form in the sperm whale accompanied with growth. Sci Rep Whales Res Inst Tokyo 17:1–14

Noordenbos W (1905) Über die Entwicklung des Chondrocraniums der Säugetiere. Petrus Camper, ned Bijdr Anat 3:367–430

Norris KS (1964) Some problems of echolocation in cetaceans. In: Tawolga WN (ed) Marine Bioacoustics. Pergamon, New York, pp 317–336

Norris KS, Dormer KJ, Pegg J, Liese GJ (1971) The mechanism of sound production and air recycling in porpoises: a preliminary report. In: Proc 8th Ann Conf on Biol Sonar and Diving Mammals, Fremont, pp 113–129

Norris KS, Harvey GW (1972) A theory for the function of the spermacetiorgan of the sperm whale. In: Galler SR et al (eds) Animal orientation and navigation. NASA Special Publication, Belleville

Novacek M (1993a) Genes tell a new whale tale. Nature 361:298–299

Novacek M (1993b) Reflections on higher mammalian phylogenetics. J Mamm Evol 1:3–30

Oelschläger HA (1978) Erforschungsgeschichte, Morphologie und Evolution der Cetacea. Natur u Museum, Frankfurt a Main, 108:317–324

Oelschläger HA (1986a) Comparative morphology and evolution of the otic region in toothed whales (Cetacea, Mammalia). Amer J Anat 177:353–368

Oelschläger HA (1986b) Tympanohyal bone in toothed whales and the formation of the tympanoperiotic complex (Mammalia: Cetacea). J Morphol 188:157–165

Oelschläger HA (1987) Pakicetus inachus and the origin of whales and dolphins (Mammalia: Cetacea). Gegenbaurs morphol Jahrb 133:673–685

Oelschläger HA (1989) Early development of the olfactory and terminalis systems in baleen whales. Brain Behav Evol 34:171–183

Oelschläger HA (1990) Evolutionary morphology and accoustics in the dolphin skull. In: Thomas J and Kastelein R (eds) Sensory Abilities of Cetacenas. Plenum, New York, pp 137–157

Oelschläger HA (1992) Development of the olfactory and terminalis systems in whales and dolphins. Chemical Signals in Vertebrates VI. In: Doty RL and Müller-Schwarze D (eds) Chemical signals in Vertebrates VI, Plenum, New York

Oelschläger HA, Buhl EH (1985a) Development and rudimentation of the peripheral olfactory system in the harbor porpoise Phocoena phocoena (Mammalia: Cetacea). J Morphol 184:351–360

Oelschläger HA, Buhl EH (1985b) Occurrence of an olfactory bulb in the early development of the harbor porpoise Phocoena phocoena. In: Duncker HR and Fleischer (eds) Vertebrate Morphologie. Fortschr Zool 30:695–698

Oelschläger HA, Buhl EH, Dann JF (1987) Development of the nervus terminalis in mammals including toothed whales and humans. In: Demski LS and Schwanzel-Fukuda M (eds) The terminal nerve (Nervus terminalis). Structure, Function and Evolution. Ann NY Acad Sci 519

Olmstead M (1911) Das Primordialcranium eines Hundeembryos. Ein Beitrag zur Morphologie des Säugetierschädels. Anat Hefte 43:339–374

Paliza OG (1964) Desarollo morfologico del feto de cachalote, Physeter catodon L. Bol Inst Invest Recurs Marin La Punta (Callao, Peru) 1:139–1960

Parry DA (1949) The swimming of whales and a discussion of Gray's paradox. J Exp Biol 26:24–34

Pilleri G (1970) Observations on the behavior of Platanista gangetica in the Indus and Brahmaputra Rivers. Invest on Cetacea, Bern 2:27–60

Pilleri G (1989) Balaenoptera siberi, ein neuer spätmiozäner Bartenwal aus der Pisco-Formation Perus. In: Pilleri G (ed) Beiträge zur Paläontologie der Cetacea Perus. Hirnanatomisches Institut Ostermundigen, Bern

Pilleri G (1990) Adaptation to water and the evolution of echolocation in the Cetacea. Ethology, Ecology and Evolution 2:135–163

Pilleri G, Gihr M (1969) On the anatomy and biometry of Stenella styx Gray and Delphinus delphis L (Cetacea: Delphiniidae). Invest on Cetacea, Bern 1:15–73

Pilleri G, Gihr M (1971) Differences observed in the skulls of Platanista gangetica (Roxburgh, 1801) and indi (Blyth, 1859). Invest on Cetacea, Bern 3:13–21

Pilleri G, Gihr M (1972a) Contributions to the knowledge of the cetaceans of Pakistan with particular references to the genera Neomeris, Sousa, Delphinus and Tursiops and description of a new Chinese porpoise (Neomeris asiaeorientalis). Invest on Cetacea, Bern 4:107–162

Pilleri G, Gihr M (1972b) Burmeister's porpoise Phocaena spinipinnis, Burmeister, 1865, off the Punta del Diablo, Uruguay. Invest on Cetacea, Bern 4:163–172

Pilleri G, Gihr M, Kraus C (1970) Feeding behaviour of the Gangetic dolphin, Platanista gangetica, in captivity. Invest on Cetacea, Bern 2:69–73

Pilleri G, Gihr M, Kraus C (1971) Further observations on the behaviour of Platanista indi in captivity. Invest on Cetacea, Bern 3:34–42

Pilleri G, Gihr M, Kraus C (1983) Near field, interference far field and rostrum structure in the echolocation system of cetaceans. Invest on Cetacea, Bern 15:11–101

Pilleri G, Gihr M, Kraus C (1987) The organ of hearing in cetacea. I. Recent species. Invest on Cetacea, Bern 22:43–125

Pouchet G, Beauregard H (1885) Note sur l'organe des spermaceti. C r Séanc Soc Biol 2:342–344

Pouchet G, Beauregard H (1889) Recherches sur le cachalot: anatomie I-V. Nouvelles archives du Mus hist nat Paris 3:1–96

Praderi R, Cave AJE (1994) Meckelian ossicle incidence in Pontoporia blainvillei. Invest on Cetacea, Bern 25:299–307

Purves PE (1966) Anatomical and experimental observations on the cetacean sonar system. Anim Sonar Syst Biol Bionics 1:197–270

Purves P, Pilleri G (1978) The functional anatomy and general biology of Pseudorca crassidens (Owen) with a review of the hydrodynamics and acoustics in Cetacea. Invest on Cetacea, Bern 9:66–227

Purves PE, Pilleri G (1983) Echolocation in whales and dolphins. Academic, London

Rajtová V (1972a) Morphogenesis des Chondrocraniums beim Meerschweinchen (Cavia porcellus L.). Anat Anz 130:176–206

Rajtová V (1972b) Über die Morphogenesis des Chondrocraniums beim Goldhamster (Mesocricetus auratus WRTH.). Anat Anz 130:207–221

Raven HC, Gregory WK (1933) The spermaceti organ and nasal passages of the sperm whale (Physeter catodon) and other Odontocetes. Amer Mus Nov 677:1–18

Reinbach W (1952) Zur Entwicklung des Primordialcraniums von Dasypus novemcinctus Linné (Tatusia novemcincta Lesson) I. Z Morph Anthrop 44:375–444
Reinbach W (1953) Zur Entwicklung des Primordialcraniums von Dasypus novemcinctus Linné (Tatusia novemcincta Lesson) II. Z Morph Anthrop 45:1–72
Reinbach W (1955) Das Cranium eines Embryos des Gürteltieres Zaedys minutus (65 mm Sch.-St.). Morph Jb 95:79–141
Reinbach W (1963) Das Cranium eines menschlichen Feten von 93 mm Sch.-St.-Länge. Zur Morphologie des Cranium älterer menschlicher Feten I. Z Anat Entwickl Gesch 124:1–50
Reinhard W (1958) Das Cranium eines 33 mm langen Embryos des Mantelpavians, Papio hamadryas L. Z Anat Entwickl Gesch 120:427–455
Reynolds SH (1897) The vertebrate skeleton (2nd edn, 1913). University Press, Cambridge
Reysenbach de Haan FW (1957) Hearing in whales. Acta Oto-Laryngologica, Suppl 134:1–114
Rice DW (1967) Cetacea. In: Anderson S and Jones JK Jr (eds) Recent mammals of the world; a synopsis of families. Ronald Press, New York
Rice DW (1977) A list of the marine mammals of the world (3rd edn). NOAA Technical Report NMFS SSRF-711, Washington DC, pp 1–15
Rice DW (1984) Ceaceans. In: Anderson S and Jones JK (eds) Orders and families of recent mammals of the world. New York, pp 447–490
Ridewood WG (1922) Observations on the skull in fetal specimens of whale of the genera Megaptera and Balaenoptera. Phil Trans R Soc Lond B 211:209–272
Ridgway SH, Demski LS, Bullock TH, Schwanzel-Fukuda M (1987) The terminal nerve in odontocete cetaceans. Ann NY Acad Sci 519:201–212
Romer AS (1956) Osteology of the reptiles. Chicago
Rothausen K (1968) Die systematische Stellung der europäischen Squalodontidae (Odontoceti, Mamm.). Paläont Z 42:83–104
Roux GH (1947) The cranial development of certain Ethiopian "Insectivors" and it's bearing an the mutual affinities of the group. Acta Zoologica 28:165–379
Schenkkan EJ (1973) On the comparative anatomy and functions of the nasal tract in Odontocetes (Mammalia, Cetacea). Bijdragen tot de Dierkunde 43:127–159
Schenkkan EJ, Purves PE (1973) The comparative anatomy of the nasal tract and the spermaceti organ in the Physeteridae (Mammalia, Odontoceti). Bijdragen tot de Dierkunde 43:93–112
Schliemann H (1966) Zur Morphologie und Entwicklung des Craniums von Canis lupus f. familiaris L. Morph Jb 109:501–603
Schliemann H (1987) The solum nasi of the mammalian chondrocranium with special references to the Carnivora. In: Kuhn H-J and Zeller U (eds) Morphogenesis of the Mammalian Skull, Mammalia Depicta 13, Beihefte z Z f Säugetierkunde, pp 91–103
Schliemann H (1991) On the development of the mammalian cranial skeleton. In: Pfeifer G (ed) Craniofacial abnormalities and clefts of the lip, alveolus and palate, 4th Hamburg Internat Symposium, Thieme Verlag Stuttgart, pp 124–130
Schneider R (1955) Zur Entwicklung des Chondrocraniums der Gattung Bradypus. Morph Jb 95:209–301
Schreiber K (1916) Zur Entwicklungsgeschichte des Walschädels. Das Primordialcranium eines Embryo von Globicephala melas (13,3 cm). Zool Jb Anat 39:201–236
Schrenk F (1989) Zur Schädelentwicklung von Ctenodactylus gundi (Rothmann 1776) (Mammalia: Rodentia). Cour Forsch-Inst Senckenberg, Frankfurt/Main, 108:1–241
Schulte HW (1916) Monographs of the Pacific Cetacea. II. The Sei whale (Balaenoptera borealis Lesson): 2. The anatomy of a foetus of Balaenoptera borealis. Mem Am Mus Nat Hist NS 1(6):389–502
Schultz AH (1935) The nasal cartilages in higher primates. Amer J Phys Anthropol 20:205–212
Seydel O (1899) Über Entwicklungsvorgänge an der Nasenhöhle und am Mundhöhlendache von Echidna, nebst Beiträgen zur Morphologie des peripheren Geruchsorgans und des Gaumens der Wirbeltiere. Semon Zool Forschungsreisen in Australien, 3 (II,1):445–532
Simon K-H (1981) Beitrag zur Craniogenese von Callithrix jacchus L. (Platyrrhini, Primates). Cour Forsch-Inst Senckenberg 45:1–106
Sinclair JG (1951) The terminal olfactory complex in the porpoise. Tex J Sci 3:251
Sinclair JG (1966) The olfactory complex of dolphin embryos. Tex Rep Biol Med 24:426–431

Slabý O (1960) Die frühe Morphogenese der Nasenkapsel beim Menschen. Acta Anat 42:105–175
Slabý O (1986) Incidence of ancestral characters in the morphogenesis of specialized structures. Folia Morphol 34:100–108
Slabý O (1987a) Contribution to a knowledge of the morphogenesis of the nasal apparatus of the dromedary (Camelus dromedarius L.). Folia Morphol 35:400–409
Slabý O (1987b) Recapitulation changes in the course of evolution. Folia Morphol 35:410–419
Slabý O (1990a) Morphogenesis of the nasal apparatus of the red deer (Cervus elaphus L.). Folia Morphol 38:212–224
Slabý O (1990b) Contribution to knowledge of the morphogenesis of the nasal apparatus of the fallow deer Dama dama L. Folia Morphol 38:408–417
Slabý O (1991) The nasal apparatus of the red squirrel (Sciurus vulgaris L.) embryo at the stage of the fully formed chondrocranium. Functional and Developmental Morphology, Academia Prague, 1(2):61–80
Slijper EJ (1936) Die Cetaceen, vergleichend-anatomisch und systematisch. Capita Zoologica 7:1–590
Slijper EJ (1962) Whales. Hutchinson, London
Spatz WB (1963) Zur Entwicklung des Nasenskeletts bei Säugetieren. Naturwissenschaften 50:454
Spatz WB (1964) Beitrag zur Kenntnis der Ontogenese des Craniums von Tupaia glis (Diard 1820). Morph Jb 106:321–416
Spöndli H (1846) Über den Primordialschädel der Säugetiere und des Menschen. Inaug dissertation Zürich
Stadtmüller F (1936) Kranium und Visceralskelett der Säugetiere. In: Bolk L; Göppert E; Kallius E; Lubosch W (eds) Handbuch der vergleichenden Anatomie der Wirbeltiere, Bd. IV, Berlin-Wien, pp 839–1016
Starck D (1941) Zur Morphologie des Primordialcraniums von Manis javanica. Desm. Morph Jb 86:1–122
Starck D (1943) Ein Beitrag zur Kenntnis der Morphologie und Entwicklungsgeschichte des Chiropterencraniums. Das Chondrocranium von Pteropus semindus. Z Anat Entwickl Gesch 112:588–633
Starck D (1953) Morphologische Untersuchungen am Kopf der Säugetiere, insbesondere der Prosimier. Ein Beitrag zum Problem des Formwandels des Säugetierschädels. Z wiss Zool 157:169–212
Starck D (1960) Das Cranium eines Schimpansenfetus (Pan troglodytes, Blumenbach 1799) von 71 mm Scheitel-Steißlänge, nebst Bemerkungen über die Körperform von Schimpansenfeten. Morph Jb 100:559–647
Starck D (1962) Das Cranium von Propithecus spec. (Prosimiae, Lemuriformes, Indriidae) (Beiträge zur Kenntnis des Primatencraniums III). Bibl primat (Schultz-Festschrift) 1:163–196
Starck D (1967) Le crâne des Mammifères. In: Grassé W (eds) Traité de Zoologie, P P, Paris, Bd XVI (1):405–549
Starck D (1975a) Embryologie. Thieme, Stuttgart
Starck D (1975b) The development of the chondrocranium in primates. In: Luckett WP, Szalay FS (eds) Phylogeny of the primates. A. multidisciplinary approach. New York, London, pp 127–155
Starck D (1979) Vergleichende Anatomie der Wirbeltiere. Bds. 2, Das Skeletsystem. Springer, Berlin Heidelberg New York
Starck D (1982) Zur Kenntnis der Nase und des Nasenskelettes von Tarsius (Mammalia, Primates, Tarsioidea). Zool Garten NF 52 (5/6):289–304
Starck D (1984) The nasal cavity and nasal skeleton of Tarsius. In: Niemitz C (ed) Biology of Tarsiers. Fischer, Stuttgart, pp 275–290
Struthers PH (1927) The prenatal skull of the Canadian Porcupine (Erethizon dorsatus). J Morph Physiol 44:127–216
Sturm H (1937) Die Entwicklung des praecerebralen Nasenskeletts beim Schwein (Sus scrofa domestica) und beim Rind (Bos taurus). Z wiss Zool 149:161–220
Szalay FS (1969) The Hapalodectinae and a phylogeny of the Mesonychidae (Mammalia, Condylarthra). Amer Mus Nov 2361:26
Taylor MA (1987a) A reinterpretation of ichthyosaur swimming and buoyancy. Palaeontology, London, 30/3:531–535
Taylor MA (1987b) Reptiles that took on the sea. New Scientist, London 11:46–51

Terry RJ (1917) The primordial cranium of the cat. J Morph 29:281–433
Thenius E (1969) Stammesgeschichte der Säugetiere (einschließlich der Hominiden). In: Helmcke JG; Starck G; Wermuth H (eds) Handbuch der Zoologie, 48 Lief, Berlin 8(2):369–722
Thenius E (1987) Stammesgeschichte. In: Grzimek B (eds) Grzimeks Enzyklopädie, Säugetiere, Kindler 4:354–357
Timm S (1987) Zur Morphologie und Entwicklung des Craniums von Flis silvestris f catus Linné 1758 – ein Beitrag zur vergleichenden Anatomie der Carnivora. Gegenbaurs Morphol Jahrb 133:3–6, 411–467, 687–729, 793–835
Timm S, Schliemann H, Sturm C-D (1989) On the morphology and development of the side wall of the cranium in fissiped cranivora. In: Splechtna H and Hilgers H (eds) Progress in Zoology, Vol 35, Trends in Vertebrate Morphology, Fischer Verlag, Stuttgart, pp 401–505
Toeplitz CH (1920) Bau und Entwicklung des Knorpelschädels von Didelphis marsupialis. Zoologica 27:1–84
Tyson E (1680) Phocaena or the anatomy of a porpess. London
Van Beneden PJ, Gervais FLP (1868–80) Ostéographie des Cétacés, vivants et fossiles. Bertrand, Paris
Van Bree PJH (1979) On the validity of the subspecies Cephalorhynchus hectori bicolor. Invest on Cetacea, Bern 4:182–186
Vandebroek G (1964) Recherches sur l'origine des mammifères. Ann Soc Roy Zool Belg 94:117–160
Van Valen L (1968) Monophyly or diphyly in the origin of whales. Evolution 22:37–41
Voit M (1909) Das Primordialcranium des Kaninchens unter Berücksichtigung der Deckknochen. Anat Hefte 38:425–616
Von Baer KE (1826) Die Nase der Cetaceen, erläutert durch Untersuchung der Nase des Braunfisches (Delphinus phocoena). Isis 1826:811–840
Von Michalkovics V (1899) Nasenhöhle und Jacobsonsches Organ. Anat H 11:1–107
Wade M (1990) A review of the Australian Cretaceous longipinnate ichthyosaur Platypterygius (Ichthyosauria, Ichthyopterygia). Mem Queensland Mus, Brisbane 28/1:115–137
Walker RA (1964) Mammals of the world. John Hopkins Press, Baltimore
Warich S (1986) Beitrag zur ontogenetischen Entwicklung der Regio ethmoidalis von Galago senegalensis (Primates). Med dissertation Univ Frankfurt
Watson L (1981) Whales of the world. Hutchinson, London
Wen IC (1930) Ontogeny and phylogeny of the nasal cartilages in primates. Contrib Embryol 130:111–134
Wiman C (1946) Über Ichthyosaurier und Wale. Senckenbergiana 27:1–11
Wünsch D (1975) Zur Kenntnis der Entwicklung des Craniums des Koboldmaki, Tarsius bancanus borneanus, Horsfield, 1821 (Beiträge zur Kenntnis des Primaten-Craniums Nr. IV). Selbstverlag des Zentrums der Morphologie, Frankfurt/Main
Yablokov AV (1964) Convergence or parallelism in the evolution of cetaceans. Paleont Zh 1:97–106 (Transl from Russ in Int Geol Rev 7:146–1468, 1965)
Youssef EH (1966) The Chondrocranium of the Albino Rat. Acta Anat 64:586–617
Zeller U (1983) Zur Ontogenese und Morphologie des Craniums von Tupaia belangeri (Tupaiidae, Scandentia, Mammalia). Dissertation Med Fakultät, Univ Göttingen
Zeller U (1985) Die Ontogenese und Morphologie des Fenestra rotunda und des Aquaeductus cochleae von Tupaia und anderen Säugern. Morph Jb 131:179–204
Zeller U (1987) Morphogenesis of the mammalian skull with special references to Tupaia. Kuhn HJ und Zeller U (eds) Mammalia depicta, Beih Z Säugetierkunde, S 17–50
Zeller U (1988) The lamina cribrosa of Ornithorhynchus (Monotremata, Mammalia). Anat Embryol 178:513–519
Zeller U (1989) Die Entwicklung und Morphologie des Schädels von Ornithorhynchus anatinus (Mammalia: Prototheria: Monotremata). Abh Senckenberg Naturforsch Ges 545:1–188
Zeller U, Wible JR, Elsner M (1993) New ontogenetic evidence on the septomaxilla of Tamandua and Choloepus (Mammalia, Xenarthra), with a reevaluation of the homology of the mammalian septomaxilla. J Mamm Evol 1(1):31–46
Zimmermann GW (1976) Beitrag zur Morphogenese des Osteo- und Chondrocraniums bei Cavia porcellus (Linnaeus). Med dissertation Univ Frankfurt
Zuckerkandl E (1909) Über den Jacobsonschen Knorpel und die Ossification des Pflugscharbeines. Sitzber Akad Wiss Wien, Math-Nat Klasse 117:493574

Printing and binding: Druckerei Triltsch, Würzburg